Controlling
Environmental
Risks from Chemicals

Controlling Environmental Risks from Chemicals
Principles and Practice

Peter Calow

Department of Animal & Plant Sciences,
University of Sheffield, UK

JOHN WILEY & SONS

Chichester • New York • Weinheim • Brisbane • Singapore • Toronto

363.7384
C 16c

Other Wiley Editorial Offices

John Wiley & Sons, Inc., 605 Third Avenue,
New York, NY 10158-0012, USA

VCH Verlagsgesellschaft mbH, Pappelallee 3,
D-69469 Weinheim, Germany

Jacaranda Wiley Ltd, 33 Park Road, Milton,
Queensland 4064, Australia

John Wiley & Sons (Asia) Pte Ltd, 2 Clementi Loop #02-01,
Jin Xing Distripark, Singapore 129809

John Wiley & Sons (Canada) Ltd, 22 Worcester Road,
Rexdale, Ontario M9W 1L1, Canada

Library of Congress Cataloging-in-Publication Data

Calow, Peter.
    Controlling environmental risks from chemicals : principles and practice / by Peter Calow.
    p.   cm.
    Includes bibliographical references and index.
    ISBN 0-471-96995-8
    1. Pollution — Environmental aspects.   2. Pollution — Risk assessment.   I. Title.
    TD195.C45C35   1997   96-39526
    363.738'4—dc21   CIP

British Library Cataloguing in Publication Data

A catalogue record for this book is available from the British Library

ISBN 0-471-96995-8

Typeset in 10/12pt Ehrhardt from the author's disks by Dobbie Typesetting, Tavistock, Devon
Printed and bound in Great Britain by Biddles Ltd, Guildford and King's Lynn
This book is printed on acid-free paper responsibly manufactured from sustainable forestation, for which at least two trees are planted for each one used for paper production.

Contents

PREFACE   ix

GLOSSARY   xi

1 PRINCIPLES   1
   1.1 Introduction   1
   1.2 An initial classification of "legislation"   4
   1.3 Basis of controls   5
   1.4 Caution with precaution   6
   1.5 Chemical controls and sustainable development   8
   1.6 Résumé   8
   1.7 Further reading   8

2 THE SCIENCE   11
   2.1 The basic questions   11
   2.2 What is it that we are trying to protect?   12
   2.3 What is actually measured?   15
         2.3.1 Retrospective approach   17
         2.3.2 Predictive approach   19
   2.4 Risk assessment   22
   2.5 Coping with the large numbers: QSARS and prioritisation   27
   2.6 Assessing the risks of accidents   30
   2.7 Résumé   31
   2.8 Further reading   31

3 RISK MANAGEMENT METHODOLOGY   33
   3.1 The forms risk management can take   33
   3.2 Risk management by command and control   34
   3.3 The involvement of private action   35
         3.3.1 Nuisance   36
         3.3.2 Negligence   37
         3.3.3 Trespass   37

3.4   Economic instruments   38
   3.4.1   Charges   38
   3.4.2   Creation of markets   39
   3.4.3   Subsidies   39
   3.4.4   Deposit/refund schemes   39
   3.4.5   Enforcement incentives   39
3.5   Registers, "audits" and labels   40
   3.5.1   Registers   40
   3.5.2   Audits   40
   3.5.3   Labels   41
3.6   Voluntary agreements   42
3.7   Résumé   43
3.8   Further reading   43

4   EUROPEAN AND UK AXIS   45
4.1   EC, EU and environment   45
4.2   The institutions of the EU   46
4.3   Involvement of institutions in legislation   47
4.4   The legal instruments   51
4.5   Evolution of environmental policy   52
4.6   Résumé   54
4.7   Further reading   55

5   SPECIFIC LEGISLATION IN BRITAIN AND EUROPE   57
5.1   Introduction   57
5.2   Information gathering and presentation as labels   58
   5.2.1   Labelling, and new chemicals   59
   5.2.2   Dangerous preparations   68
   5.2.3   Pesticides   68
   5.2.4   Existing substances   69
5.3   Controlling legislation   72
   5.3.1   Pollution controls for (largely) point sources in aquatic systems   72
   5.3.2   Pollutant controls for (largely) point source emissions to
       atmosphere   78
   5.3.3   Controls on marketing and use; a more distributed problem   81
   5.3.4   Waste   84
   5.3.5   Major accidents   84
5.4   How it fits together   85
5.5   Towards more integration   86
5.6   Résumé   89
5.7   Further reading   89

6   US LEGISLATION WITH SOME NOTES ON CANADA
    AND THE REST OF THE WORLD   91
    6.1   Introduction   91
    6.2   The framework   91
          6.2.1   Pollution prevention   93
    6.3   Toxic Substances Control Act (TSCA)   93
          6.3.1   Distinction between new and existing chemicals 94
          6.3.2   New chemicals   94
          6.3.3   Existing substances   95
    6.4   The Federal Insecticide, Fungicide and Rodenticide Act (FIFRA)   97
    6.5   The Clean Air Act (CAA)   98
    6.6   The Clean Water Act (CWA)   99
    6.7   Comprehensive Environmental Response Compensation and Liability
          Act (CERCLA or Superfund) 1980 and the Superfund Amendments
          and Reauthorisation Act (SARA) 1986   100
    6.8   Other relevant legislation   101
          6.8.1   Federal Food, Drug and Cosmetic Act (FFDCA)   101
          6.8.2   Occupational Safety and Health Act (OSHA)   101
          6.8.3   Safe Drinking Water Act (SDWA)   101
          6.8.4   Resource Conservation and Recovery Act (RCRA)   102
          6.8.5   Hazardous Materials Transportation Act (HMTA)   102
          6.8.6   Consumer Product Safety Act (CPSA)   102
          6.8.7   Federal Hazardous Substances Act (FHSA)   102
          6.8.8   National Environmental Policy Act (NEPA)   102
          6.8.9   Wildlife and Wilderness Acts   103
    6.9   Résumé and comparisons with European legislation   103
    6.10 Canada   104
    6.11 Rest of the world   104
    6.12 Further reading   104

7   INTERNATIONAL ORGANISATIONS AND PROGRAMMES   105
    7.1   Introduction   105
    7.2   UN Programmes   105
          7.2.1   UNEP   106
          7.2.2   ILO   107
          7.2.3   FAO   107
          7.2.4   WHO   107
          7.2.5   IMO   108
          7.2.6   Other UN organisations   108
    7.3   Organisation for Economic Cooperation and Development
          (OECD) Chemicals Programme   109
    7.4   Council of Europe   111
    7.5   Paris Commission (PARCOM)   111
    7.6   North Sea Conference   111

7.7	European Community/Union	112
7.8	NGOs	112
7.9	Standards organisations	113
7.10	Résumé	115
7.11	Further reading	115

8   THE FUTURE   117

8.1	The current position	117
8.2	State of the environment and résumé of trends	117
8.3	Some challenges for the science	121
8.4	A basis for integration	122
8.5	Assessing and managing risks from major accidents	123
8.6	Taking benefits as well as costs into account	124
8.7	Risk management options	127
8.8	Résumé	129
8.9	Further reading	129

Preface

This is a deliberately slim book that aims to bring information, and hopefully insight, on how industrial chemicals are controlled for the sake of environmental protection to those who need an overview. It is not intended to be detailed in any of the many disciplines that inevitably impinge on this area: chemistry, ecology, ecotoxicology, environmental sciences, economics, law, risk assessment, risk management, etc., etc. Each chapter ends with a list of key texts that will act as stepping stones into the detailed literature.

The focus here, then, is on the principles and practices of how the environmental (by which I mean ecological) hazards and risks associated with industrial chemicals are recognised and how this information is used in their management. I have attempted to bring out the core principles such as addressing the question – What are we trying to protect when we talk of ecological systems? – while at the same time giving some detail on how these issues are incorporated into the legislation. The emphasis in these latter terms has been on Europe, but, in part recognition of the fact that "chemical pollution knows no national boundaries", chapters are also devoted in turn to North American and International involvement.

As readers I have had in mind: students on courses that emphasise the environmental sciences but are fuzzy on the legislative contexts; regulators who are well-versed in the legislation but unclear on the basic principles of ecology, ecotoxicology, risk assessment and so on; and people in industry who because of everyday pressures of the legislation have too little time to appreciate its basis and context.

Despite its size, therefore, the book covers a lot of territory – literally – from UK to EU to UN, and also from science to legislation. This would not have been possible without advice and comments from many. I am particularly grateful to Dr Pat Murphy who gave useful advice from an EU perspective, and Dr Chris Lee and Dr Maurice Zeeman from a US perspective. Simon Ball, who sadly died before the book was published, gave much appreciated advice on the legal issues. As usual, though, I have to be responsible for what has ultimately emerged.

There are two final caveats. Legislation is a very dynamic entity, subject to revision by governments, refinement by regulators and interpretation by the Courts. This is especially so of environmental legislation, and so what you have in this edition is a snapshot, hopefully in focus at the time of writing, but possibly blurring with time. Because of this, and the frankly personal interpretation that I have put on the principles and their developments, especially as considered in the final chapter, readers would be unwise to make commercial or legal decisions solely on the basis of what is contained in this book.

I hope, nevertheless, that what I have written provides perspective, insight and provocation in what is a rather complex, but crucial area; and I should be pleased to receive any comments, positive or negative, from readers on how well these aims have been achieved and, indeed, on any aspects of the book.

Peter Calow
October, 1996

Glossary

The following is a comprehensive listing of acronyms and abbreviations used in the text

AFNOR	Association Française de normalisation (France)
ANSI	American National Standards Institute (USA)
AQMA	Air quality management areas (UK)
AQS	Air quality standards (UK)
ASTM	American Society for Testing and Materials
ATP	Adaptation to Technical Progress (Instrument associated with EC legislation)
BAT	Best available techniques
BEO	Best environmental option
BPEO	Best practicable environmental option
BSI	British Standards Institution (UK)
CAA	Clean Air Act (US)
CEC	Commission of the EC
CEI	Comitato elettrotecnico italiano (Italy)
CEN	Comité Européen de Normalisation
CEPA	Canadian Environmental Protection Act
CEQ	Council of Environmental Quality (US)
CERCLA	Comprehensive Environmental Response Compensation and Liability Act (US – Superfund)
CFC	Chlorofluorocarbon
CFR	Code of Federal Regulations (US)
CIA	Chemical Industries Association
CICADs	Concise international chemical assessment document(s) (UN)
CIMAH	Control of Industrial Major Accident Hazards
CIS	International Centre on Occupational Safety and Health Information
CL	Critical concentration (level)
COPA	Control of Pollution Act (UK)
CPSA	Consumer Product Safety Act (US)
CPSC	Consumer Product Safety Commission (US)
CWA	Clean Water Act (US)
DG	Directorate General (of the CEC)

DIN	Deutsches Institut für Normung e.V. (Germany)
DoE	Department of Environment (UK and US)
DoJ	Department of Justice (US)
DOT	Department of Transport (UK and US)
DS	Dansk Standardiseringrad (Denmark)
E(L)(I)C$_{50}$	Concentration causing specified effect (lethality, inhibition) of 50% population. Might be expressed as dose, e.g. LD$_{50}$.
EC	European Community
ECETOC	European Centre for Ecotoxicology and Toxicology of Chemicals
EEA	European Environment Agency (EU)
EEB	European Environmental Bureau
EEC	European Economic Community(ies)
EHD	Estimated human dose
EINECS	European Inventory of Existing (Commercial) Chemical Substances
EIS	Environmental impact statement
ELINCS	European List of Notified New Chemical Substances
ELOT	Hellenic Organisation for Standardisation (Greece)
EPA	Environmental Protection Act 1990 (UK) [see also USEPA]
EQO	Environmental quality objective
EQS	Environmental quality standard
EU	European Union
FAO	Food and Agriculture Organisation (UN)
FDA	Federal Drug Administration (US)
FEPCA	Federal Environment Pesticide Control Act (US)
FFDCA	Federal Food, Drug and Cosmetics Act (US)
FHSA	Federal Hazardous Substances Act (US)
FIFRA	Federal Insecticide, Fungicide and Rodenticide Act (US)
FPS	Formal group on priority setting
GLP	Good laboratory practice(s) (OECD/EU)
HEDSET	Harmonised electronic data set (associated with EC Existing Substances Regulation)
HMIP	Her Majesty's Inspectorate of Pollution (now merged into the Environment Agency of England and Wales with NRA).
HMTA	Hazardous Materials Transportation Act (US)
HPV	High production volume(s)
HSE	Health and Safety Executive (UK)
IBN	Institut belge de normalisation (Belgisch Instituut voor Normalisatic) (Belgium)
ICCA	International Council of Chemical Associations
ICEF	International Federation of Chemical, Energy and General Workers Union
IGC	Intergovernmental Conference (of the EU)
IIRS	Institute for Industrial Research and Standards (Ireland)

ILO	International Labour Organisation
IMCO	International Maritime Consultation Organisation (now IMO)
IMO	International Maritime Organisation
IOC	Intergovernmental Oceanographic Commission (UN)
IP(P)C	Integrated pollution (prevention) control
IPCS	International Programme on Chemical Safety (WHO)
IPS	Informal working group on priority setting
IRPTC	International Register of Potentially Toxic Chemicals (UN)
ISO	International Standards Organisation
IUPAC	International Union of Pure and Applied Chemistry
K_{OW}	Octanol-water partition coefficient
LCA	Lifecycle assessment
LNC	Levels of no concern
LOEC	Lowest concentration in a gradient causing a significant difference in the observed trait from controls
MAB	Man and Biosphere Programme (UN)
NAAQS	National ambient air quality standards (US)
NEEC	Not entailing excessive costs
NEPA	National Environmental Policy Act (US)
NGO	Nongovernmental organisation
NIHHS	Notification of Installation Handling of Hazardous Substances
NNI	Nederlands Normalisatie Instituut (Netherlands)
NOEC	Highest concentration on a gradient not causing a significant difference in the observed trait from controls
NPDES	National Pollutant Discharge Elimination System (CWA)
NPL	National Priority List (CERCLA)
NRA	National Rivers Authority (now part of the Environment Agency of England and Wales with HMIP)
OECD	Organisation for Economic Co-operation and Development
OJ	*Official Journal of the EU*
OSHA	Occupational Safety and Health Act (US)
OSPARCOM	Oslo Paris Commission
PARCOM	Paris Commission
PCP	Pentachlorophenol
PEC	Predicted environmental concentration
PIACT	International Programme for the Improvement of Working Conditions and Environment
PIC	Prior informed consent
PMN	Premanufacture notice (TSCA)
PNEC	Predicted no effect concentration
PNOAEL	Predicted no observed adverse effect level
(Q)PAR	(Quantitative) property activity relationship
(Q)SAR	(Quantitative) structure activity relationship

RCRA	Resource Conservation and Recovery Act (US)
RIA	Regulatory impact assessment (under TSCA)
RM	Risk management process (TSCA)
SARA	Superfund Amendments and Reauthorisation Act
SCOPE	Scientific Committee on Problems of the Environment
SDD	Sustainable development debate
SDWA	Safe Drinking Water Act (US)
SEA	Single European Act (EC)
SIDS	Screening information dataset (OECD)
SOP	Standard operating procedure
TBT	Tributyltin
TIE	Toxicity identification evaluation
TRI	Toxic release inventory(ies)
TSCA	Toxic Substances Control Act (US)
TSMP	Toxic Substances Management Policy (Canada)
TU	Treaty of Union (Maastricht Treaty) (EU)
UN	United Nations
UNEP	United Nations Environment Programme
UNESCO	United Nations Educational Scientific and Cultural Organisation
UNI	Ente nazionale italiano di unificazione (Italy)
U/S	Uses and substitutes reports (under TSCA)
UNIDO	United Nations Industrial Development Organisation
USEPA	United States [of America] Environmental Protection Agency
VOC	Volatile organic compound
WAAG	World-at-a-glance (USEPA document)
WHO	World Health Organisation
WWF	World Wide Fund for Nature

1

Principles

1.1 INTRODUCTION

At least since the invention of fire, more than 100 000 years ago, humans have been introducing novel chemicals into the environment. This must have increased in pace with the development of metal-working approximately 5000 years ago; but especially over the last 200 years following the Industrial Revolution. Figure 1.1, not to be taken too literally, illustrates this pace of change.

Now there are more than 10 million synthetic chemicals of which 100 000 (1%) are known to be traded internationally. It is estimated that less than 10% of these make up more than 90% of the total production of chemicals. About 2000 new substances are being notified per year in both the European Union[1] and the USA. More facts and figures are given in Table 1.1.

These chemicals are produced because we use and need them. Either directly or indirectly they have made enormous contributions to human quality of life in terms of freedom from disease, famine, poverty and simply convenience of living. Yet it is a long-recognised fact that in sufficient quantity, any chemical, even the so-called natural ones, can cause problems for human health and the environment. There is therefore a need for regulation.

This book aims to describe these controls, especially in so far as they are applied for the protection of the natural environment and with particular emphasis on how they are being applied in Europe. The "synthetic chemicals" are often classified by use: those used for industrial purposes; those used in agriculture, for example as plant protection products and fertilisers; and those used pharmaceutically. Here the main emphasis will be on industrial chemicals. However, the concern will not only be with the details of specific controls, but also with the philosophical, scientific and policy basis of the controls that can be applied to synthetic chemicals of any kind.

---

[1] The abbreviation EU will be used to refer to the collection of Member States; EC (European Community) will be used to refer to its legislation (see Chapter 4).

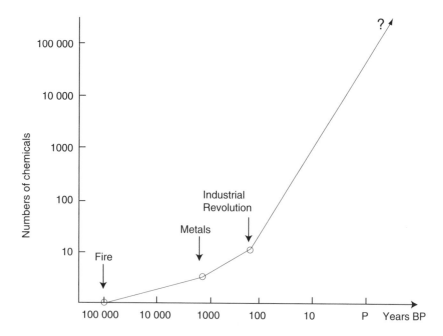

Figure 1.1   Number of chemicals generated by human activity, and potential environmental contaminants

This introductory chapter begins by describing some general principles that will be elaborated further in Chapter 3. Much of the legislation involves an increasing reliance on scientific evidence of hazard and risk. Consequently, this methodology and the assumptions associated with it will be addressed in Chapters 2 and 3. Chapter 4 outlines the legislative institutions and mechanisms in Europe and the UK, paying particular attention to how they interact, prior to descriptions of specific legislation in Chapter 5. The latter will be centred on the EU, as the initiator of much of UK legislation. Despite subsidiarity, this is how it ought to be in a context in which the subjects, chemicals, are traded internationally and have their impacts globally. Some attention will be paid to the "rest of the world" but especially to North American legislation in Chapter 6, and to international bodies and conventions concerned with chemical controls in Chapter 7. Finally, and somewhat speculatively, the book will return to the question of attempting to balance risks with benefits, particularly from the perspective of the development of sustainable and globally practicable chemical controls policy. More generally this final chapter suggests scenarios for future action in terms of protecting the environment from possible harm from commercial chemicals.

Table 1.1   Some statistics on chemical production

Statistics	Source
1. 10m. chemicals registered scientifically; c. 75% of these 10m. mentioned only once in the scientific literature	IRPTC (1991). 10 million chemical substances registered. *IRPTC Bulletin*, **10**, 9
2. Approximately 100 000 existing chemicals in commercial use in Europe	European Inventory of Existing Commercial Chemical Substances (EINECS) (Annex to *Official Journal of the European Communities*, **53**, June 1990; C146A)
3. Of which c. 20 000 likely to be dangerous but only c. 2500 have been officially classified	Murphy, P. (1992). The systematic evaluation of existing chemicals in the European Community. In *Chemicals Control in the European Community*, pp. 43–45, Royal Society of Chemistry, Cambridge
4. 1500 new substances notified to EU by industry between 1981 and 1991	Murphy, P. & Rigat, P. (1994). *The Notification of New Substances in the European Union*. Official Publication of the EC 1994
5. US Toxic Substances Control Act (TSCA) Chemical Substance Inventory of existing chemicals lists > 70 000	Worobec, M. D. & Hogue, C. (1992). *Toxic Substances Controls Guide* (second edition). Washington DC
6. Now between 1000 and 2000 notifications a year of new chemicals in the EU and USA	
7. By contrast to a relatively modest production before World War II, in the late 1970s production was c. 400m. tonnes p. a. In 1989 it was estimated that the production of synthetic materials had increased 350-fold since 1940. Probably <10% of chemicals account for >90% total production	Tolba, M. K. & El-Kholy, O. A. (Eds) (1992). *The World Environment 1972–1992*. Chapman & Hall, London, on behalf of UNEP
8. Nearly 6 trillion pounds (c. 2.7 trillion kg) of industrial chemicals are produced or imported into the USA p.a.	Inform (1995). *Toxic Watch 1995*. New York.
9. Worldwide c. 5 billion pounds (2.3 billion kg) of 1600 different pesticides are applied annually; <0.1% of these reach their target pests	Pimental, D. (1995). Amounts of pesticides reaching target pests: environmental impacts and ethics. *Journal of Agriculture & Environmental Ethics*, **8**, 17–29

1.2   AN INITIAL CLASSIFICATION OF "LEGISLATION"

In reviewing chemicals legislation it is convenient to make a distinction between that which is concerned with **information gathering** (prioritising, classifying and labelling) and that which is concerned with **control** (Figure 1.2). This distinction is also in line with a separation that is often advocated between assessment procedures, that should be as objective as possible, and risk management that will also involve socio-political judgements. However, these distinctions are blurred, because the collection of information itself, particularly when distilled into a classification that is made public, e.g. through a label, may exert constraints. And what kind of information is collected and how it is assessed may well depend, as we shall see, on what value society puts on various ecological entities.

Historically, within the controlling legislation there has also been something of a separation between controls applied to **emissions** and those applied to the **distribution, marketing and use** of chemicals and articles containing them. The former are sometimes referred to as pollution control and are often concentrated on waste streams and point sources such as stacks and pipes; the latter are sometimes referred to simply as chemical legislation and address a much more distributed problem. Again, though, the distinction between these two classes of control is not too sharp; for example, judgements about controls on marketing and use could include considerations on waste emissions.

Controls, of course, need not always be applied directly as regulatory instruments. Increasingly, market (economic) instruments are being harnessed in environmental protection in general. They come in many forms and we have already touched on one: the use of labels to influence public opinion and consumer choice. Another form is the application of differential taxation favouring more environmentally friendly products and making less favourable ones more expensive. A good example of this kind of instrument and its effects is the differential taxation on leaded as compared with unleaded petrol in the EU, that has rapidly shifted consumption to the lead-free petrol. A more thorough analysis of market instruments will be given in Chapter 3.

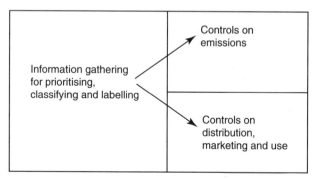

Figure 1.2   Types of chemical controlling legislation

1.3   BASIS OF CONTROLS

Another way of classifying controls is in terms of the quality and quantity of information upon which they are based. Historically this legislation was piecemeal and reactive, with the burden of proof being in a sense with the polluted, i.e. when pollution was seen to occur it was controlled. This, for example, was the basis of the origin of the British Alkali Acts that were the vanguard in pollution control legislation in Britain. But now the ethos has shifted to the obviously more sound approach of prevention rather than cure and so increasingly controls are more proactive. At the same time, there has been recognition that the environment is so complex that it is proving hard to rapidly obtain a thorough scientific understanding of it and all the consequences of chemical contaminants within it. Yet the problems that we are now creating are so profound and potentially serious that we cannot necessarily wait for a balanced scientific view. There has therefore been an increasing emphasis on the precautionary principle, where protecting the environment is given the benefit of any doubt; so the absence of evidence or scientific understanding precisely linking effects with chemicals is not necessarily being allowed to stand in the way of the imposition of controls.

As far as chemicals go, application of the precautionary principle can take several forms (see Figure 1.3). An extreme form would be to take every opportunity to restrict our use and dependence on all chemicals, given that all can cause problems in certain circumstances. But this would be unrealistic **and** undesirable (see below). Some chemicals are more problematic than others because they have adverse effects even at low concentrations. Being more precise about the identification of these chemicals involves the use of scientific evidence.

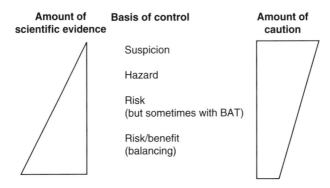

Figure 1.3   Classification of the basis of controls on chemicals in terms of the balance between the involvement of scientific evidence and the precautionary principle

As a first step in this direction the approach may be unsystematic; any evidence that chemicals are likely to have some adverse effects at low concentrations is used as a basis for considering controls. This is a **suspicion-based** approach. The so-called Black and Grey priority lists of chemicals that are part of EC legislation to protect the aquatic environment against pollution from effluent discharges were initially compiled largely in this way (see p. 73).

A more systematic approach is to base judgements on evidence about the potential of chemicals to cause significant problems from laboratory tests that investigate the relationship between concentration and effects. This is a **hazard-based** approach. With this approach hazard levels are defined, often somewhat arbitrarily, and if breached the substance is either banned or singled out for special treatment. This is the way that inclusion within the Black list has subsequently been confirmed, or otherwise, as part of the EC procedure (see pp. 75–76).

A further level of sophistication involves basing judgements about controls on the likelihood of the potential to cause harm (hazard) being realised in realistic environmental circumstances. This involves estimating the probability that a concentration of a chemical that is likely to cause harm will be exceeded in the environment. This is a **risk-based** approach, and is increasingly being used in the development of control measures on, for example, the marketing and use of chemical substances in the EU (pp. 85–86). Here the options for control are more than just banning; having identified levels below which effects are unlikely, various kinds of restrictions can be required to ensure these are not exceeded. It can still involve the application of precaution by introducing safety margins into the estimates of expected environmental concentrations and effect concentrations. Further precaution can also be applied by requiring that the allowable levels defined by risk assessment are treated as maxima or minima to be bettered by using so-called best available techniques and technology (BAT). This is a **technology-based** approach.

1.4   CAUTION WITH PRECAUTION

The precautionary principle has at least two facets: (1) insufficient scientific evidence should not be allowed to stand in the way of implementing controls especially when there is prima-facie evidence of serious consequences; and (2) the natural world is such a complex place that science can never hope to identify all the possible effects of contaminants, so even with some scientific understanding, we should always build safety margins into our controls. These are apparently such reasonable viewpoints that it is somewhat difficult to represent a counterview. Yet there are some good reasons for not pushing these precautionary principles too far.

For one thing, by so doing important benefits to human health and welfare may be denied as a result of the unnecessary restriction of availability and use of a chemical. True, with more than 100 000 synthetic chemicals available there ought to be functional substitutes, but these are not necessarily without risks themselves and are often less well studied and characterised than the common chemicals. Moreover, it can be argued that the amount of overall resource (time and effort as well as money) available for environmental protection is ultimately limited, so using more than is necessary in certain areas means that less will be available for protection in others and this will be to the detriment of environmental protection in general.

Not only, then, is there a need to base controls on as much scientific evidence as possible, but also to balance the needs for control with their relative effectiveness and with the benefits that derive from the use of the chemicals. These are the **balancing-based approaches** listed in Figure 1.3. They come in various forms, for example:

- Do the benefits from reduced risks outweigh the lost benefits associated with the use of the chemicals in terms of, for instance, protection from disease, improved food production, quality of life? This is a consideration that is now required in contemplating controls of existing chemicals in both the EU and the USA (p. 125).
- Do the benefits from reduced risks outweigh the costs of compliance with the controls? Here, for instance, there may be a question of comparing the investment in new technology with returns in terms of extra environmental protection, leading to the idea that BAT (best available techniques) should be balanced with NEEC (not entailing excessive costs) as is required in Part I of the UK Environmental Protection Act 1990 (p. 80).
- Do the benefits from reduced risks outweigh the costs of enforcement? This clearly involves a consideration of the costs of monitoring and prosecutions in the event of failure to comply – but may also involve a consideration of the costs of risk assessment procedures themselves. The latter are potentially very considerable, especially taking into account the costs of carrying out tests, so the view here would be that only those chemicals for which there is prima-facie evidence for concern should be subject to the full rigour of an assessment. This is a **prioritisation-based** approach that is a first step in the risk assessment of existing chemicals in both the EU (p. 71) and the USA (p. 96). It essentially balances a perception of potential risks with the costs of risk analysis.

Balancing-based approaches stand somewhat in tension with the precautionary approach, in that they accept the possibility of allowing some pollution. There is, therefore, a tacit acceptance that it is reasonable to put the health of some and/or the well-being of some ecological systems at risk for the greater good. This raises some profound ethical questions; for not all, with this approach, are necessarily being treated equally under the law. We shall address the principles of **hazard identification** and **risk assessment** further in the next chapter and return to the **balancing approaches** in the final chapter.

1.5   CHEMICAL CONTROLS AND SUSTAINABLE DEVELOPMENT

All this has to be put within the context of the **sustainable development debate** (SDD): an exploration of the view that pathways of economic development can be found that leave future generations no worse off (at least) in terms of environmental holdings than we have been. Economic development is equated with the accumulation of capital; but the SDD has made explicit that this should include human capital (the usual definition), social capital (related to education, health and welfare) and natural (environmental/ecological) capital. Chemical production undoubtedly contributes to the accumulation of human capital but at the expense of natural capital – because it uses raw materials and potentially creates pollution. At the same time, however, it can have very positive effects on social capital (through enhancing food yield, health and general welfare) and this in turn can have knock-on, positive effects to natural capital, for satiated, healthy and well-educated people are more likely to take environmental protection seriously than the hungry, ill and poorly educated. The aim of chemical controls ought to be to optimise this precarious balance; something that is now at the heart of EU thinking about programmes (section 4.5) and UN involvement (section 7.2).

1.6   RÉSUMÉ

- We produce many chemicals, some in large quantities, because they are useful.
- At the same time they can have adverse effects on human health and the environment, so regulation is required.
- Increasingly the regulation is being carried out on the basis of scientific understanding of the relationship between chemicals and their effects in the environment, and some account is being taken not only of the risks posed but also the social benefits gained from the production and use of substances.
- But there is still an inclination to apply caution in interpreting the evidence and being prepared to act without scientific understanding when there is a perception of serious problems.
- However, the precautionary principle should not be overused, because that in itself can lead to suboptimal environmental protection.

1.7   FURTHER READING

Francis, B. M. (1994). *Toxic Substances in the Environment*. John Wiley & Sons Inc., New York.
O'Riordan, T. & Cameron, J. (1994). *Interpreting the Precautionary Principle*. Cameron & May, London.

Pearce, D., Markandya, A. & Barbier, E. B. (1989). *Blueprint for a Green Economy.* Earthscan Publishers Ltd, London.

Richardson, M. L. (1988). *Risk Assessment of Chemicals in the Environment.* Royal Society of Chemistry, London.

Richardson, M. L. (1992). *Risk Management of Chemicals.* Royal Society of Chemistry, London.

2

The science

2.1 THE BASIC QUESTIONS

There are two basic and somewhat different questions concerning the control of chemicals:

1. To what extent are chemical substances likely to have adverse effects on the environment?
2. To what extent are chemical substances having adverse effects on the environment?

The first question is the kind that might be asked about new chemicals prior to them being manufactured, marketed or released. It is predictive; it ought to involve risk assessment (see Chapter 1). The second question is the kind that can be asked globally, about for example chlorofluorocarbons (CFCs) interacting with the ozone layer, or locally about the impact of an effluent that a factory is releasing into a local river. It is retrospective and often involves actual environmental assessment. Of course, though distinguishable, both exercises are not sharply distinct for, given our lack of understanding of complex environmental problems, retrospective monitoring ought to provide some indication of whether we have got the answers to the predictive questions right.

Deciding how much harm might be (or is being) done is a matter for science; it can be done by reference to critical analysis and carefully controlled observation. It is the province of ecotoxicology. This part of the process, it is often argued, should be separated from the more subjective decisions about whether or not the harm is important and what should be done about it. Yet once again the distinction is not so clear-cut because in ecotoxicology management decisions have to be taken about what is to be protected before any measurement can be made and this clearly has implications for what and how measurements are made. These management decisions not only have a scientific content – in defining natural systems and their natural states – but also a socio-political one in deciding what it is about nature that society values and wants to protect. In the rest of this chapter we address the scientific issues before going on to a consideration of some of the issues for risk management in the next chapter.

2.2   WHAT IS IT THAT WE ARE TRYING TO PROTECT?

Environment is a broad term and not easily defined. To say that everything around us is environment while strictly correct is not very useful. The French origin – *environner* – implies "encirclement" and support, and is usually taken to mean the physical and chemical surroundings of life; the air, earth and water in which organisms live. Yet the composition of the atmosphere, the formation and composition of soils and sediments, and the concentration of substances in waters all crucially depend upon the action of organisms; and, of course, there are important interactions between one organism and another (see below). So "environment" properly includes both the physical, chemical **and** biological surroundings in which organisms live. The thin, life-containing skin around the planet that extends up into the atmosphere and down into the soil and seas is referred to technically as the biosphere and is a synonym for the global environment.

The concept of the environment as "supporting" has been taken by some as far as implying that this is directed and responsive. The "earth mother" provides just those conditions that are appropriate for life and actively maintains these, in the same way that control systems such as thermostats maintain a constant temperature in the face of fluctuation in ambient conditions. This is the basis of the well-known Gaia theory, named after the Earth goddess of the ancient Greeks. It suffers from a number of difficulties, whereas there is an acceptable scientific theory to explain how organisms come to have attributes suitable for supporting their existence and a capacity to actively maintain these in the face of disturbance – through evolution by natural selection – there is no such theory for the biosphere. For natural selection to operate, there needs to be competition for limited resources and genetic continuity, so those attributes associated with success are perpetuated. The biosphere does not have these organismic characters; it is not a superorganism. Nor does it have the organisation that is necessary for active thermostat-like controls; there are no control centres and no specific feedback channels. Individual organisms, on the other hand, do have neural and endocrine control centres and channels.

This can be taken further. Organisms interact with others to form collective groups: of the **same species** through reproductive and sometimes social encounters and competition to form populations; with **different species** as food, feeder, competitor, etc. to form communities. Within the latter there are fluxes of energy (ultimately from sunlight) and material (ultimately from the physical surrounds) along food–feeder channels. These ecological systems are known as ecosystems. They have describable structures (species composition; biodiversity) and processes (fluxes; see Figure 2.1) that can equilibrate and demonstrate more or less stability in the sense of resisting disturbance and recovering afterwards. But again these are not actively regulated.

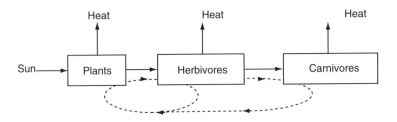

Figure 2.1 Simplified food chain showing energy flow (——) and material cycles (- - -). It could be grass ⇒ cattle ⇒ human, or tree ⇒ insect ⇒ bird. Usually the patterns are much more complex

Species do not cooperate for the common good. Indeed, quite the contrary. Species that selfishly use resources for their own perpetuation will be favoured in the short term because they leave more offspring with the same selfish tendencies even if this spells disaster for the system in the long term by overexploitation. Just as the biosphere cannot be treated as an adapted and adaptable superorganism, the same is the case for individual ecosystems. The recent tendency to talk in terms of ecosystem health, in so far as it implies properties normally associated with actively regulated organisms, is therefore somewhat misleading.

Because organisms adapt to the environment and within ecosystems and not the other way round, it follows that it is not possible to define what is a normal environment, except by reference to organisms. Hence this will be in terms of conditions to which the organisms in question have adapted. Indeed, given time, organisms can often adapt, by natural selection, to abnormal circumstances; as is well illustrated by the evolution of resistance to pesticides and tolerance to pollutants in natural populations of plants and animals. What we are usually concerned with, therefore, is the extent to which contaminants so alter the environment that conditions diverge from those to which organisms are adapted sufficiently to impair their survival and reproduction. More generally, we may be concerned with "average" conditions for the survival and persistence of organisms in a particular ecosystem or region.

For similar reasons, it also follows that it is not possible to be very precise about the properties of "normal" ecosystems. They certainly have properties and involve processes but these are not directed towards specific adaptational ends. Organisms within ecosystems do not cooperate for some common good though they interact "functionally" in a common system. The relationship between structure and process is a somewhat open ecological question. That there has to be some relationship is beyond doubt – by definition ecosystems could not function without species – but the tightness of the relationship is open to question. Certainly ecologists have demonstrated in simplified laboratory ecosystems that more diverse systems involving annual plants and their associated insects and soil fauna are more productive than species-poor systems. And in experimental plots, involving prairie plants, it has been demonstrated that more diverse communities used nutrients more efficiently and hence with more production than less diverse plots. On the other hand, there is evidence that in more complex communities, there can sometimes be much functional redundancy in that species can be replaced and even removed with no replacement without effect on the functional attributes.

In yet other ecological research it has been shown that, notwithstanding the possibility of redundancy, some individual species can have very important effects on the other species that live together with them by more subtle means. Thus if animals with very particular feeding habits are excluded, those organisms that constitute their food might be allowed to increase in abundance with knock-on effects for organisms on which they feed and with which they compete, thus altering the whole community. For example, ecologists have excluded mussel-eating starfish from stretches of seashore. This caused the mussels to grow in abundance and exclude barnacles and seaweeds, making the manipulated shore look quite different from the natural shore. The starfish are thus known as keystone species. The extent to which ecosystems in general are structured by keystone species is unknown; but it is unlikely that all of them will be.

Another related question concerns the extent to which ecosystems can be said to be stable and how this is related to biodiversity, and affected by contaminants. Of course, and as already noted, ecosystems can potentially develop equilibrium states of species components and the fluxes of energy and matter within them, that are more or less capable of being disturbed and able to recover afterwards. But this is not by active feedback; it is simply what happens to parts that interact in systems. Thus pendulums have stable states to which they tend after deflection according to Newtonian principles; similarly, mixtures of reactive chemicals can tend towards stable equilibria maintained by often subtle feedback systems based on the laws of diffusion and reaction. But there are no active or intentional goals and controls here. It is also possible, in principle, to define ecosystem condition in terms of the characteristics that affect stability and consider how pollutants affect these. However, one of the surprises that has come out of modern ecology is that some lines of evidence suggest that ecological simplicity rather than complexity promotes this kind of stability.

This has to be hedged in various ways. One particular qualification, for example, is that ecosystem stability has to be context dependent, i.e. more complexity, and hence less intrinsic stability, is possible in stable environments and vice versa. This is one reason why tropical terrestrial ecosystems are so diverse; because the weather and temperature regimes to which they are exposed are not very variable, or at least their variability is predictable. Rivers and streams, on the other hand, suffer much variability in flow regimes – that is, they are unpredictable – and they tend to contain low-diversity communities with short food chains. Now it seems intuitively obvious, and there is some evidence for this, that pollution will tend to reduce species diversity and shorten food chains and this could, somewhat paradoxically, lead to more stable systems! Clearly this viewpoint is not undisputed; but the main point is that a deterioration in ecosystem condition might be associated with changes from structural or functional "norms", but these changes are not unquestionably in the direction of reduced ability to resist further changes or to recover from them.

Moreover there is another problem with the equilibrium/stability approach to ecosystems and this is that stable equilibria may not exist in nature. In dynamic environments such as the rivers and streams, there is probably a continual opening up of ecological space by the washing away of populations and communities. What invades will depend upon what is present locally in undisturbed stretches of neighbouring rivers and streams, their mobility, and chance. Initial invaders may be later replaced by incoming competitors. So there is a complex dynamic patchwork in which it may be hard to predict the course of events. This means that there may be no such things as ecological norms.

A different way of considering ecosystem condition is in terms of: what is in it for us? Increasingly it is recognised that ecosystems provide services in terms of, for example, supporting atmospheric quality, influencing climate, supply of biomass and water, providing recreational opportunities and aesthetic returns. Therefore, we can judge ecosystems in terms of the extent to which they are supplying these services and we can design management strategies to protect them in this way. It then becomes of importance to consider to what extent the provision of services depends upon the existence of particular characteristics of structure and process within ecosystems. Most of the services referred to above depend upon processes: the cycles of matter (water, carbon, nitrogen, etc.) and the fluxes of energy. So questions about tightness of coupling between structure (biodiversity) and process and the properties that promote the stability of processes become of central importance.

2.3   WHAT IS ACTUALLY MEASURED?

In principle, then, we ought to have a clear focus on what we want to protect and why. This will have both a social dimension and a scientific dimension. Society needs to understand what ecological services it receives and prioritise them. Science needs to explore what factors affect services and how they can be operationalised for measuring and predicting relevant effects, when the questions raised above about the relationship between structure and process become of importance. A distinction is sometimes made between assessment endpoints (e.g. levels of primary production, fertility of soil, yield of fish) and measurement endpoints (diversity indices, results from toxicological tests). The assessment endpoints ought to be guided by the kinds of consideration that were raised in the last section. Thus if our aim is to protect services and there is a tight relationship between structure and process, species effects (survival/reproduction) are important operational measures; but if there is poor coupling, possibly we ought to measure processes directly (decomposition rates, productivity). On the other hand, if keystone species occur, it is on them and their responses that measurement should focus.

A related issue concerns the level of organisation at which we should be making our observations. For biological systems this ranges from organisms "up" to collective groups such as populations, communities and ecosystems, and "down" into their physiology, cellular structure and function, molecular structure and function and so on (Figure 2.2). Toxic chemicals generally have their first effects at suborganism levels; to denature important biomolecules and to disrupt physiological processes. Measurements on effects here can usually be carried out more quickly (and hence cheaply) and with much more experimental control and rigour than measurements on populations and especially communities. On the other hand, not all suborganismic effects necessarily translate into organismic ones, nor organismic ones into population ones and so on. Thus exposure to metals can induce the synthesis of a specific class of identifiable proteins called metallothionenes. These can be used as specific indicators of metal exposure, but because their function appears to be to neutralise the toxic effects of metals they may not signal higher-level effects. They are part of the protection machinery of organisms. Similarly mortality of a few organisms may have no impact on population size, since the loss may mean that the per capita supply of resources, e.g. food, is increased for those remaining, so increasing their survival chances and/or reproductive outputs. We have already noted that loss of a few species within a community may not have much effect on its general level of diversity and functioning.

Looking at it the other way round, "top down" rather than "bottom up", it is also possible to miss effects with components of ecological systems by observing them in

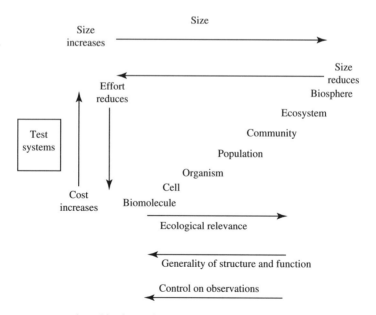

Figure 2.2   Hierarchy of biological organisation with associated features

isolation. Thus the effect of a toxicant on a population may depend not so much on how it affects individuals in that population directly but as a result of differential action, i.e. how it affects the balance in predator–prey or competitive interactions between species.

Clearly in making our measurements we ought to have a clear idea of what we want to assess and hence to protect. Because of a combination of fuzzy thinking and the complexity of ecological systems this is often not the case. Rather what we measure is driven by habit and convenience. Often what we do, at least implicitly if not explicitly, is try to define assessment endpoints as no ecological change and design measurements accordingly. The rationale (usually not articulated) is that this is precautionary; no change, by definition, should protect whatever it is that delivers appropriate services. Albeit a reasonable stance, this is probably unrealistic for two reasons: ecosystems are naturally dynamic and economic development is probably impossible without some ecological effect.

In any event, observing change in particular systems, for example as a result of the release of effluent by a factory into a particular river (retrospective analysis) ought to be more straightforward than forecasting what changes might take place in the environment at large as a result, for example, of the production of a new chemical product (predictive analysis). Hence retrospective programmes ought to be more straightforward than predictive ones. This is generally the case, but both exercises are not without their problems.

2.3.1   Retrospective approach

This can either be involved in assessing the general quality of the environment (surveillance) or specific sites and problems (monitoring). In either case, and no matter whether the physico-chemical or biological aspects are being considered, the most important challenges are in establishing a baseline (tantamount to defining what constitutes normality), picking up abnormal changes from normal ones and ascribing blame, i.e. that the change is caused by some contamination.

For example, arguments about the importance of the greenhouse effect have stemmed from difficulties in deciding if the small increase in surface temperature ($< 1\,°C$) over the last century could be attributed to a now undisputed large ($c.\ 95\%$) increase in accumulation of greenhouse gases in the atmosphere or to natural processes. There are reasons for believing that the two are not necessarily linked but that, due to the buffering effects of the oceans, the impact of the greenhouse gases on temperature has simply been put off. Models have been designed to make predictions about what is likely to happen in the future, but cannot do this with great confidence because we do not understand how complications such as cloud cover and the storage effects of the oceans are likely to distort the greenhouse effects.

Arguments about acidification from "acid rain" have similarly revolved around difficulties of picking up subtle changes in acid (pH) values in natural systems. It is difficult to make comparisons in contemporary systems, but here analyses of lake sediments have allowed ecologists to look back into the past. Cores of sediment are taken from lake bottoms and these contain pollen grains and diatom cases accumulated in the past. Different species show different tolerances for high or low acidity so species identification plus chemical analysis can show up trends in acidification. In both Europe and North America this kind of approach provides evidence of increased acidification with increasing industrialisation.

The problems of picking up changes and ascribing them with confidence to specific causes are exacerbated in biological assessment since here it is unclear as to what is most appropriately observed (see above). Much work has been done on flowing waters (rivers and streams) because these have been used to take wastes away from sewage works and factories. Bottom-dwelling invertebrate animals are frequently used as "indicators" of disturbance, basically because they are relatively easy to sample and identify. In Britain much store has been put in particular species that are thought to be especially sensitive or tolerant to pollution and these have been used as indicators or so-called biotic indices. Unfortunately, it is looking increasingly likely that no one species will be universally sensitive or tolerant to all pollutants. The rank order of sensitivities that has been used in Britain probably reflects the order of sensitivity to organic loading from sewage works, and this does not necessarily hold for other pollutants.

The general tactic, no matter what is being measured, is to *compare* sites suspected of pollution with clean sites. This is the only way of proceeding if, as indicated in the last section, we cannot predict "norms" from first principles. The reference sites can either be upstream of a point of pollution, another similar site in a similar river/stream or, less usually, the site under consideration *before* treatment, or in some more sophisticated designs all of these together. A significant difference is taken to indicate disturbance following the "no change" criterion discussed above. But this immediately raises the question as to whether the reference sites really are representative of the conditions in the study site, had there been no "treatment". And this, of course, is not easy to justify, particularly in systems such as flowing waters where, as already noted, there can be considerable spatial **and** temporal dynamism. Confidence can be improved by using more than one reference site and making observations over an extended time period, but there can still be room for doubt.

Another approach uses models to generate expectations. For example, in Britain the Institute of Freshwater Ecology has developed a model that predicts species lists for particular sites. This was established by taking invertebrate samples and measuring a large number of physical and chemical variables in an initial number of river stretches that were designated as clean. Sophisticated multivariate analysis was then applied to consider if a predictive model could be constructed on the basis of as small a number of physico-chemical variables as possible. It turned out that reasonable predictions were possible from only five variables. There are still some potential problems with this kind of approach – e.g. in defining clean sites in the first place, and recognising that a correlation between variables and species does not necessarily imply causality, because there may be hidden variables correlated with the ones measured that are actually controlling diversity. However, it does present the possibility of establishing a baseline against which divergence in type, number and abundance of species can be judged which would be strengthened if a mechanistic link could be established between dependent and independent variables in the model. As an aside, it also suggests that there may be more pattern to even the most dynamic systems than was suggested in the last section.

All the above focus on structural aspects of ecosystems, but this might miss other kinds of changes that signal harm. A different approach is taken by the US Environmental Protection Agency (USEPA). Under the Clean Water Act (p. 99) the USEPA is required to "evaluate, restore and maintain the physical **and biological integrity** of the Nation's Waters". Biological integrity is judged in relation to the state of a suite of criteria based on both the structural and functional attributes of resident fish and bottom-dwelling invertebrate communities using baseline criteria obtained from reference sites within a defined ecoregion – i.e. a relatively homogeneous area in terms of geography, hydrology, land use, etc. A problem again, though, is that the attributes measured are chosen more for convenience than on the basis of any understanding of how they relate to ecosystem integrity.

Finally, a weakness with all these after-the-fact correlations is that in the rich *mélange* of variables that constitute the natural world some that might be instrumental in a change go unnoticed. Of course, this will depend, to some extent, on magnitude and timing. Thus, there may be little doubt about cause and effect immediately after a major disaster, such as an oil spill – but as time proceeds the significance of ecological aspects of shore organisms and whether they are caused by lingering pollution become more obscure. One method that tries to address this kind of problem is to expose controlled test systems to samples from a putatively contaminated site in a systematic way – the so-called toxicity identification evaluation (TIE) approaches. But these then are subject to the criticism that the effects (or their absence) may not be relevant.

In retrospective studies it is a fact of life that uncertainties about effects and causes conspire to confuse. Akin to medical epidemiology and forensic work, weight of evidence, from a variety of sources, and expert judgement are likely to be more prominent than probability statements on likelihood of effect and cause of the kind that we are used to from scientific experiments.

2.3.2 Predictive approach

In principle, if we understand what effects are of importance we could test chemicals against appropriate systems to predict likely consequences of their release. The best that we can usually do, however, is to get an impression of hazard by reference to responses of standard tests. From this we attempt to obtain an impression of risk from defining threshold concentrations below which effects are unlikely and comparing these with likely environmental concentrations. Risk assessment will be addressed in more detail in the next section. As a preliminary to that, some general comments are made here about the tests themselves. This is the subject area of ecotoxicology.

The tests can be classified in various ways: involving single or many species; exposure to high concentrations over short periods of time relative to the life-span of the test organisms (acute tests) or at lower concentrations over longer time periods (chronic tests). Single species are used more frequently than multi-species, because they are easier to handle. Since keystone species may not occur in all ecosystems, especially loosely structured ones such as rivers and streams, it is not generally possible to justify choice of test organisms in these terms. Nor does there seem to be any prospect of finding a generally sensitive species. Instead, test organisms are usually selected for convenience; and basically this means convenience of keeping them in the laboratory. Clearly tests that focus on individuals presume that what is being measured have "bottom up" relevance and/or there are no appreciable "top down" effects of the kind already discussed above.

In tests it is important to recognise that effects become more likely as exposure concentration increases (known as a dose (concentration) response relationship) and at particular concentrations as exposure time increases. It seems reasonable to expect that effect at a concentration increases with time until with more time there is no further effect. The effect should therefore be measured at this critical time beyond which no further effect is observed. This will change, of course, with concentration and the effect being measured.

Rarely are such detailed analyses possible, particularly in routine testing circumstances. Instead acute tests usually involve measuring mortality or related characters over predetermined short periods (days/hours) and a series of high concentrations. Chronic tests measure more subtle, but nevertheless ecologically relevant responses, often in individual growth and reproduction, over longer periods (often days/weeks) and a series of lower concentrations. Examples of some typical tests are given in Box 2.1.

The results of acute tests are commonly expressed as the concentration that has a defined effect on a given fraction of the population. The concentration affecting 50% is most often used as the index because most of the population clusters around this value and hence it can be estimated with most statistical confidence. In contrast the results of chronic tests are often expressed in terms of no *observed* effect concentrations (NOEC) which is the highest concentration at which no effect was detected – or lowest *observed* effect concentrations (LOEC) – which is the lowest concentration (next to the NOEC) for which an effect was recorded. "Observed" is included in the NOEC partially in recognition of the fact that negatives cannot be proved and in both the NOEC and LOEC in recognition of the fact that their values depend importantly on the concentration intervals used in the test; there may be a lower concentration nearer to the NOEC, which produces an effect but where no measurements were made, and similarly there may be higher NOECs. Another problem with both NOECs and LOECs is that they assume a threshold – i.e. sharp jump – between

Box 2.1   Some typical ecotoxicological tests used in testing chemicals in Europe

The following are part of so-called base-set requirements used for testing new substances likely to be marketed at between 1 and 10 tonne p.a. (see p. 61). The original concept was that this base set represented a simple aquatic food chain from primary producer, to herbivore and finally to predatory fish.

- **Acute toxicity for fish** – Fish are exposed to test substances added to water at a range of concentrations for a minimum of 48 h, but preferably 96 h. Choice of species is at the discretion of the laboratory – but should be from a single stock of similar length and age, in good health and free from any apparent malformation. Mortalities are recorded and an LC_{50} (concentration that kills 50% of the population) computed.
- **Acute toxicity for *Daphnia*** – Daphnids are "water fleas" that are supposed to be representatives of the plankton. Groups (at least 20) more than 6 h and less than 24 h old are exposed to a range of concentrations over at least 24 h. Immobilisation is used as an endpoint; this is a precursor of lethality. Hence, the endpoint is defined in terms of an effect rather than lethality, i.e. EC_{50}. (see Glossary)
- **Growth inhibitor test on algae** – Populations of algal cells are exposed to increasing concentrations of test substances. Cells divide and this is how populations grow. The extent to which populations have grown is assessed after 3 days and compared with controls (no chemicals). Results can be expressed in terms of that concentration inhibiting x% of normal growth (e.g. IC_{50}), (see Glossary) or having no effect relative to controls. This is a test that is relatively short for the observer but long with respect to the life-span of the test organism (undergoes several generations over 3 days). Moreover, population growth depends upon the balance of reproduction (cell division) and mortality. So though short term the test is probably more properly described as chronic.

When the quantity of chemical marketed is more than 10 tonnes p.a. so-called level 1 testing is required. This includes:

- **Prolonged toxicity study with *Daphnia magna*** (21 days) – A number (10) of daphnids is established at each of a range of concentrations. Total production of offspring is monitored over 21 days and compared with controls to give an NOEC.

Other tests called for but not as precisely defined are: test on higher plants; test on earthworms; further toxicity studies with fish.
   When the quantity of chemical marketed exceeds 100 tonnes p.a. some or all so-called level 2 tests are required. This will involve: further toxicity studies with fish; toxicity studies with birds; additional toxicity studies with other organisms.

concentrations that have or do not have effects and these may not exist. Certainly the concentration/response relationships of some toxicants follow a more continuous curve; an effect reducing continuously with reducing concentration to zero without break. Finally, because of the difficulties associated with NOECs and LOECs it is increasingly being suggested that results should be expressed in terms of effect concentrations predicted to impact a particular fraction of population or community as in the acute tests or even extrapolated mathematically to zero effects.

Note should also be made of the use of the term "concentration" rather than "dose" which is more usual in toxicology. Dose is what is delivered to the tissues either in food or by injection. In ecotoxicology we usually measure exposure related to what it is in the surrounding environment; so concentration is more appropriate. "Nominal concentration" is what is delivered, but part of this may be lost during a test due to degradation, adsorption to surfaces and organic particles in the system – e.g. food particles or sediment: so "actual concentration" is what is measured during or at the end of tests. How concentrations translate into dose is complex, depending upon uptake characteristics, and possibly also breakdown and excretory processes of the organism.

2.4 RISK ASSESSMENT

Risk is the probability that a specified harmful effect will occur. Most, if not all, processes, actions and objects have the potential to cause harm to somebody and/or something. The potential is referred to as **hazard**. Hazards are assessed through laboratory toxicity (for humans) and ecotoxicity (for ecological systems) tests and this is sometimes referred to as **hazard identification. Risk assessment** is concerned with estimating the likelihood of these hazards being realised for defined targets. Environmental and ecological risks of chemicals are therefore concerned with estimating the probability that a chemical with a particular level of production and pattern of use will have a significant adverse effect on the environment, humans in the environment and ecological systems. This therefore involves estimating the probability of **exposure** and from this the probability of adverse **effects**.

There are uncertainties at almost every step in this process. Generally, they derive from three sources:

1. That the world is fundamentally statistical – similar entities may respond differently under identical conditions as is true, for different reasons, of decaying isotopes and individual organisms – so we can define average population responses but not the response of particular individuals. This is known as **stochasticity.**
2. That we rarely have full knowledge or understanding (**ignorance**).
3. That we often make mistakes in our observations (**fallibility**).

Thus in predicting likely environmental concentrations (PECs) there are uncertainties associated with production levels (due to ignorance and fallibility), in estimating releases into the environment (again due to ignorance), in predicting how substances will partition between atmosphere, water and soil (possibly due to stochasticity; certainly due to ignorance) and in predicting levels of degradation (due to fallibility in measuring this; and again to potentially both stochasticity and ignorance). If we were able to put quantities on each of these uncertainties we would be able to predict a likely environmental concentration with a range of levels of reducing probability of occurrence. The distribution would become "tighter" the more that we can reduce all the uncertainties.

On the presumption that we have defined a target effect (cf. last section) for ecological systems – say death of a specified fraction of a population (human or other species) or extinction of a certain fraction of species in a community or even loss of a certain function within ecosystems – then there are likely to be the three usual sources of variation in our estimates of effects or no effects (predicted no effect concentration, PNEC). The first is variation between individuals or species in their responses; so-called biological variation, the stochastic effects. The second is due to fallibility in measuring the responses. The third is due to ignorance. Usually we have to reconstruct distributions of responses from very limited sampling.

Now in principle we could specify a mean environmental concentration and a mean effect level each with variances (e.g. Figure 2.3). And the risk assessment requires a judgement about the extent to which these distributions overlap. The null hypothesis is usually that exposure concentrations do not overlap effects (i.e. PEC ⩽ PNEC). There are two types of error that we can commit in making this judgement, both of

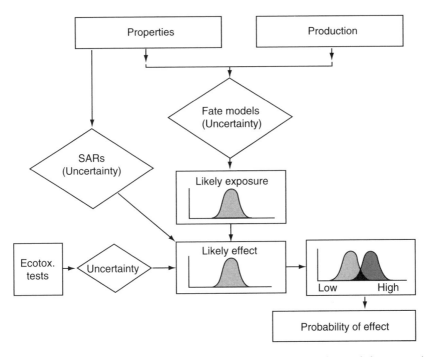

Figure 2.3   Basic principles of risk assessment. The penultimate box is left vague. The higher distribution could represent effects, with the PNEC somewhere at its lower edge; or it could be a distribution of PNECs. Alternatively, distributions could be switched with exposures being higher than effects

which increase with uncertainty, and whose likelihood can be computed from knowledge of variances. If we come to the conclusion that there is a difference where there is not one we commit what is known as a Type 1 error, i.e. wrongly reject the null hypothesis and admit false positives. A statistical probability can be computed to indicate how likely this is. Presume, on the other hand, that no difference is presumed, but that the variances are such that there is in fact a difference (PEC > PNEC). We would here commit what is known as a Type 2 error; wrongly accept the null hypothesis and admit false negatives. Statistical probabilities can be computed that indicate how likely this is. Type 1 errors are the ones most often examined by scientists; they are cautious about claiming effects that do not exist. Type 2 errors ought to be taken more seriously by ecotoxicologists because the consequences of saying that a difference does not exist, between a presumed safe level and exposure, when in fact it does, may be important.

So from this kind of analysis we should be able to predict that a certain fraction of a population, or of species within a community or even of the processing of some quantity of energy or carbon would be likely to be adversely affected with a certain level of confidence (probability). Moreover, each of the distributions concerning environmental concentrations and ecological effects could vary through time and this ought also to be reflected in the assessment. Rarely, however, do we have this thorough knowledge of levels and distributions.

Environmental concentrations might be monitored directly. But these data may not be available, as is bound to be the case for new substances; and even if they were they may only represent a "snapshot" in time and space. So we usually need to make predictions about expected concentrations at local situations from calculations usually based on particular sites or assumptions, and more regionally and globally from presumptions about average conditions often fed into generalised models of compartmentalisation. Releases have to be considered along the life-cycle of chemicals and involve the following: production volumes (calculated from industry returns, though subject to some commercial sensitivities); from formulation processes (e.g. into paints or pesticides); use (processing) by industry and public (again measured or estimated); disposal. Volumes involved at each stage multiplied by release factors give amounts lost to environment which then: partition between water, sediment/soil, atmosphere according to their physico-chemical properties; become distributed and diluted largely according to environmental conditions (such as discharge rates in rivers, wind speeds and directions); persist to an extent which depends upon the physico-chemical properties of the substance and the extent to which it is exposed to appropriate conditions (e.g. sewage treatment). PECs can, therefore, be calculated for local scenarios, usually following the precautionary principle over short timescales representing worst-case situations (e.g. those days in which maximum releases are likely in batch production or episodic use). In the EU, regional PECs represent Member States and have inputs and outputs from outside the geographical area, whereas continental PECs represent the whole EU which is presumed to be isolated and have no inputs and outputs. Both are calculated as averages over annual timescales. The results for the continental scenario are used as background for the regional, and from the regional as background for the local. In the end, therefore, it is possible to have PECs at a number of geographical scales ($PEC_{continental}$; $PEC_{regional}$; PEC_{local}) and for each of these to have PECs for each environmental compartment (PEC_{water}; $PEC_{sediment}$; PEC_{soil}; PEC_{air}).

It is as well to remember also that the "C" in PEC refers to environmental concentration. Ideally, it should relate to the effective level at the target site(s) within organisms (i.e. dose) and should involve consideration of uptake, loss and hence bioaccumulation of material. Rarely, however, do we have this information and moreover results from ecotoxicological tests are usually expressed in terms of ambient concentration. On the other hand, some kind of assessment of bioaccumulation is important to assess likely exposure through food chains. For example, prey organisms may only be exposed to low concentrations of contaminants and/or for biological reasons accumulate small amounts. But by eating many of them, predators can potentially accumulate larger quantities and be subject to adverse effects. Biomagnification of this kind has proved important in the effects of certain insecticides on insect-feeding birds. Partition of chemicals between octanol and water can give an indication of bioaccumulative potential (see next section). Consideration of simple food chain scenarios, in which it is usually presumed predators feed entirely on contaminated prey, can give some insight into these secondary poisoning events.

As indicated above, PNECs are obtained from ecotoxicological tests carried out on a few species. Because they take less effort, and hence cost less to carry out, most results are from acute tests. So caution has to be applied in calculating predicted no effect levels from these to take into account:

(A) That you can never prove a negative – no effects in one group of one species do not preclude effects at the same concentration in other groups of the same or different species;
(B) that an acute effect does not preclude an effect in another character at lower concentrations;
(C) that there is an increase in complexity in moving from simplified laboratory observations to natural ecological systems.

For these reasons assessment (sometimes referred to as safety or uncertainty) factors are applied to the laboratory results, often dividing these by 10 or 1000 depending upon the quality and quantity of the data. Even these, though, are subject to uncertainty. Indeed the factors of 10 are probably more to do with the way we count fingers on our hands than on science. Some recent exercises have attempted to introduce more rigour for ecological assessments by seeking specific, statistical correlations between acute and chronic, single- and multi-species results in a series of chemicals. But the data are necessarily limited, and a general correlation does not guarantee that it will work for particular new chemicals.

So taking into account all the above uncertainty, PNECs are best understood as concentrations *unlikely* to have ecologically relevant effects. There are, in fact, some attempts to reconstruct distributions of responses within communities from a small number of responses on representative species so that a full analysis of the kind illustrated in Figure 2.3 can be carried out. But the calculations are complex; the reconstruction is based upon presumptions about the form the distribution takes and this has yet to be substantiated experimentally; and there is a debate on representativeness – the species subset from which the reconstruction is carried out should be a random sample from natural communities, but this is rarely, if ever, the case. PECs similarly are "ballpark" estimates laced with precaution by multiple recourse to worst-case scenarios.

Box 2.2   Basic elements of risk characterisation as used in EU schemes

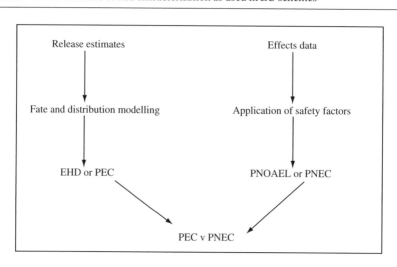

where
PNOAEL = Predicted no observed adverse effect level
PNEC   = Predicted no effect concentration
PEC   = Predicted environmental concentration
EHD   = Estimated human dose

- PNOAEL is used for humans and the level may represent a concentration (for inhaled substances) or a dose (for imbibed substances).
- PNEC is used for ecological systems.
- PEC is used for both ecological systems, and humans when the substance is inhaled or absorbed.
- EHD is used for humans when the substance is imbibed.

Therefore "risk" is often estimated by a simple comparison between the PNEC and PEC; the PEC/PNEC ratio is called a risk quotient. When PEC is greater than the PNEC (risk quotient >1) then there are grounds for concern and considering the need for managing the potential risk through the application of controls (see Box 2.2). This is certainly the approach taken in EU new and existing substances legislation; but a similar approach is adopted under the US Toxic Substances Control Act (p. 95).

Because risks cannot easily be defined precisely in ecotoxicology, and yet the exercise just described goes beyond simply assessing the potential harm associated with a substance, there has been a confusion in terminology applied to the process. "**Risk characterisation**" seems to be emerging as a descriptor for the EU, but at the end of the day what matters is not so much what terminology is used to describe a process as that what is done is clearly and unambiguously described (see section 5.2).

2.5   COPING WITH THE LARGE NUMBERS: QSARS AND PRIORITISATION

Given the large number of existing chemicals it would not be practicable, nor even possible, to carry out thorough risk assessments of them all. However, many substances, especially organic molecules, share common structural features and/or physico-chemical properties that are likely to have implications for their effects. This raises the possibility of finding relationships, often precise mathematical ones, between structure and properties and effects. These are referred to respectively as SARs and PARs (structural/property activity relationships), and when expressed in quantitative form as QSARs and QPARs. They provide short-cut devices for predicting effects for chemicals for which there are no experimental data.

Another, non-mutually exclusive, approach is to use what is known about substances to identify those in need of the most urgent attention. This is referred to as prioritisation. Many priority lists have been compiled for chemical controls and some have already been mentioned (p. 6); others will be covered in Chapter 5. Often in the past these have been compiled in a suspicion-driven way (p. 5), but more recently there have been attempts to do this more objectively, systematically and transparently. There are two main considerations: how much and which information to use and how to use the information.

On the first consideration, it is important to bear in mind that prioritisation is an initial step towards a more thorough analysis. It is therefore important that it focuses on effective indications of risk while recognising that the information available, by definition, will be limited and often incomplete. For the sake of expediency, prioritisation will usually be based on intrinsic properties, hazard rather than risk. Yet exposure is often taken into account, if only superficially, by reference to production levels, and indications of persistence deduced from tests measuring susceptibility of substances to break down, especially in the presence of microbes (biodegradation). A number of schemes use production volume as an important criterion in narrowing down the list of substances for more detailed treatment; e.g. the EU existing substances legislation uses high production volume (HPV) to determine urgency of input into a prioritisation procedure (p. 70) and the OECD use the HPV criterion to identify priorities for detailed risk assessment (p. 110). Finally, though hazard is usually judged in terms of acute responses, chronic responses can also be taken into account, as can the potential ability to accumulate in tissues as suggested by the extent to which substances partition into an organic phase, usually octanol, rather than water in standard tests (octanol:water test). The organic phase provides a simple model for biological tissues.

On the second consideration, of how to use the limited information in prioritisation, there are at least three non-mutually exclusive schemes: decision trees; scoring systems and mathematical models. One common feature in the first two of these is that they often seek to transform subjective judgements or observations from continuous distributions of data into discrete assessments as descriptors (high, medium, low) or numerical scores (usually 1 to 10). Some examples are given in Table 2.1.

Figure 2.4 gives some examples of decision trees. These provide clear pictorial representations of the logic of the decision processes. One problem, though, is that they generate priority lists, but do not rank substances within them. The scoring

Table 2.1 Numerical values for high and low thresholds for chemical property or toxic endpoints. After Hedgecott, S. (1994), in *Handbook of Ecotoxicology*, Vol. 2, pp. 368–393, Blackwell Scientific, Oxford

Property	Toxic endpoint or parameter	Unit	High	Low	Insignificant
Acute aquatic toxicity	96-h LC_{50} (fish) or 48-h EC_{50}/LC_{50} (invert)	mg/l	<1.0	>100	—
Chronic aquatic toxicity	NOEC	mg/l	<0.01	>1	—
Persistence in water	Half-life	days	>100	<10	<2
Bioaccumulation potential	Bioconcentration factor log K_{ow}	— —	>1000 >3.5	<100 <2	— —
Toxicity to higher organisms	Significant toxic effect or oral LD_{50}	mg/kg	<50	>500	—
Production and use		tonnes/year	>10 00-0	<1000	—
Solubility		mg/l	>1000	<1	—
Volatility		pascals	>0.133	<0.000133	—

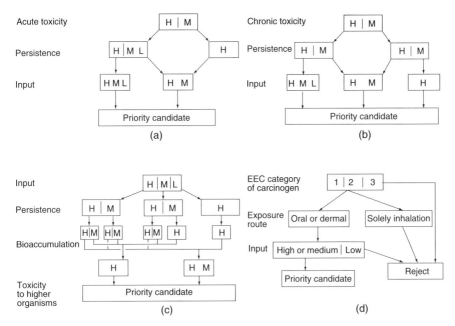

Figure 2.4 Some examples of prioritisation schemes used in generating the Red List of chemicals referred to in Chapter 5 (Table 5.1): (a) "Red List 2" short-term scenario; (b) "Red List 2" long-term scenario; (c) "Red List 2" food-chain scenario; (d) "Red List 2" carcinogenicity scenario. High (H), medium (M) and low (L) are defined in Table 2.1. Modified from Hedgecott, S. (1994) Prioritisation and standards for hazardous chemicals. In Calow, P. *Handbook of Ecotoxicology*, Vol. 2, pp. 368–393. Blackwell Science Ltd, Oxford

systems on the other hand usually combine (either by addition or multiplication) the quantified scores associated with each criterion and use the single number output to rank substances. However, care is needed because there is not necessarily direct proportionality between scores and impacts, i.e. a substance with half the score of another does not necessarily have 50% of the impact. Combination of scores can also make the logic of prioritisation decisions less clear.

Mathematical models can incorporate views about mechanisms (e.g. of dispersion and effects) with quantified parameters to yield rankings. An example of such a ranking procedure now being used to prioritise existing chemicals in the EU for more further detailed analysis (p. 71) is given in Box 2.3. Again, though, care has to be exercised in not allowing the complexity of such models to obscure the process and its aims.

A very common difficulty in all these procedures is lack of data; information gaps. One response to these is to reject substances in which they occur, but that might exclude substances that in other respects should be treated as priorities for attention. So more often gaps are filled with dummy, default values such as worst-case

Box 2.3   Automated ranking procedure developed for generating priority lists for EU existing substances legislation

---

The method is called the informal working group on priority setting (IPS) method. It is automated and uses a model.

The steps:

1. It takes information from databases supplied by producers.
2. Selection of data is based on criteria such as data from standard tests using GLP (i.e. good laboratory practice). Data on this basis is classified into: not acceptable; acceptable; preferred.
3. Model predicts exposure on basis of use, likely environmental distribution, biodegradation.
4. Model predicts effects from ecotoxicological data, possibly using QSARs to fill gaps.
5. Ranking based on product of appropriately scaled exposure and reciprocal of effects concentrations. Obviously the larger these numbers, the more the need for concern.

---

descriptors or scores or with values extrapolated from similar substances, possibly even by QSAR predictions (p. 27). The safest approach is obviously to use worst cases as defaults, but that might mean that ignorance, lack of data, drives prioritisation. Clearly there is room for judgement here, but when and how these judgements are made needs to be made clear for the sake of retaining overall transparency.

2.6   ASSESSING THE RISKS OF ACCIDENTS

The risk assessments already described are in terms of the likelihood of chemicals causing harm from intended patterns of production and use. However, also of relevance is the possibility of harm being caused from unintended, accidental releases. Risk assessment procedures can be applied to these accidental events. Here risk is defined in terms of: (1) a probability of occurrence and (2) a consequence. In a sense the second component is equivalent to the risk assessments described above: if the accident leads to release of a particular concentration of chemical (PEC) what is the likely consequence (cf. predicted no effect levels)? So it is the first part of the definition, i.e. (1) above, that is new.

What is required in defining the probability of occurrence is (a) breaking down the operation into a logically connected system of components and then (b) estimating the probability of failure of each component (often expressed as a logical fault tree). This kind of methodology has been developed to a very sophisticated level in the nuclear industry. Failure data for most mechanical systems are already available. Similar failure probabilities have been estimated for a variety of working environment scenarios and conditions, involving human subjects. All these data are largely based on past experience.

Of course, the likelihood of a failure potential (similar to hazard, p. 22) being realised and hence the extent of the risk will also depend on the extent and efficiency of protective devices and systems that are in place. These lessen the chances of the hazard being realised. So it is also necessary to systematically review each element of the process to consider information on the efficiency of protective devices and systems. This information can be summarised in the form of an event tree.

The results of these two kinds of analyses, (1) potential for failure and (2) likelihood of occurrence given the protective system in place, can be combined to give a risk profile of the operation. This then defines a risk of failure which can in principle be combined with analyses of consequences for human health, in the workplace and beyond, and ecological effects, to give a risk assessment of harmful consequences.

2.7 RÉSUMÉ

- Protecting the environment presupposes that we have targets that we aim to protect but these are not easily defined, especially for ecosystems.
- In measuring the quality of ecological systems it is usually necessary to use comparisons with supposed standard systems. Predictions about impacts are most often based on simplified tests.
- Risk assessment procedures need to take account not only of the potential of substances to cause harm (hazard) but the likelihood of that potential being realised, which requires in turn our understanding of exposure and hence fate.
- It would neither be practicable nor possible to apply a thorough risk assessment to all existing substances so methods for predicting possible effects from structure and properties are being developed, and schemes for identifying substances for priority attention are in use in a number of legislative contexts.
- Assessing the risks from chemicals released by accident requires a knowledge of system failure and the kinds of response systems that are in place to deal with them.

2.8 FURTHER READING

Calow, P. (1993). *Handbook of Ecotoxicology*, Vol. 1. Blackwell Science, Oxford.

Calow, P. (1994). *Handbook of Ecotoxicology*, Vol. 2. Blackwell Science, Oxford.

Forbes, V. E. & Forbes, T. L. (1994). *Ecotoxicology in Theory and Practice*. Chapman & Hall, London.

Kareiva, P. (1996). Diversity and sustainability on the prairie. *Nature*, **379**, 673.

Quint, D., Taylor, D. & Purchase, R. (1996). *Environmental Impact of Chemicals: Assessment and Control*. The Royal Society of Chemistry, London.

Rodricks, J. V. (1992). *Calculated Risks*. Cambridge University Press, Cambridge.

SETAC (1996). Issue 1 of *Environmental Toxicology and Chemistry*, **15**, 1–76. [For some examples of how the distributions in Figure 2.3 might be quantified.]

Suter, G. W. (1993). *Ecological Risk Assessment*. Lewis Publishers, Boca Raton.

Van Leeuwen, C. J. & Hermens, J. L. M. (1995). *Risk Assessment of Chemicals: an Introduction*. Kluwer Academic Publishers, Dordrecht/London.

3

Risk management methodology

3.1 THE FORMS RISK MANAGEMENT CAN TAKE

In Chapter 1 a distinction was made between information gathering and controls. In Chapter 2 we considered the scientific basis of information gathering. Here we describe how restrictions might be applied. Clearly in between decisions have to be made about how serious the risk is and whether or not it should be managed. Amongst other things, these considerations have to balance the risks against social benefits from the substance under consideration and we shall return to this in the final chapter.

In the rest of this chapter we make a distinction between various forms of controlling instruments:

- Command and control – the most familiar, applied on the basis of statutes by regulatory authorities representing the state.
- Private action – brought by individuals representing their own interests, but may thus bring a general benefit to society.
- Market instruments – involving consumers exerting choice in a market-place on the basis of information provided either (a) by pricing (economic instruments) or (b) more directly in registers or as labels.
- Voluntary measures – involving self-imposed controls brought into effect by those producing or marketing the substances.

It is also worth bearing in mind where the risks arise from, what general form they take, and hence where the controls can be applied. The major sources of risks are from: manufacturing and processing; storage; distribution; use (including both general and professional); and ultimate disposal. Risks might arise from normal and abnormal operations and from accidents. Finally release might be from point sources or from more distributed sources. How these characteristics relate to each other is suggested in Table 3.1.

Table 3.1   Sources of risk and methods of control

Sources	Manu-facturing	Storage	Distribution	Use	Disposal
Operating/use procedures that could cause problems					
Normal	✓	✓		✓	✓
Abnormal	✓	✓	✓	✓	✓
Accident	✓	✔	✔	✓	✓
Kind of release					
Point	✔	✔	✔	✓	✓
Distributed	✓	✓	✓	✔	✓
Appropriate control					
Command and control	✓	✓	✓	✓	✓
Market instrument	✓	?	?	✓	✓
Voluntary	✓	✓	✓	✓	✓

Bold ticks indicate where the emphasis might be in the various elements of the supply chain.

3.2   RISK MANAGEMENT BY COMMAND AND CONTROL

Controls on emissions are usually designed to eliminate pollution, and some would interpret this as a need to achieve zero emissions. However, a more realistic approach is to define pollution not by reference to the presence of a substance but in terms of its effects. Hence, it is useful to distinguish between contamination (presence of a substance not necessarily producing adverse effects) and pollution (presence of a substance producing adverse effects).

There are two possible ways that controls to prevent pollution might be specified: either in terms of what comes out of pipes and stacks (emission values) or in terms of what happens in the environment when they pass outside an appropriately defined mixing zone (so-called environmental quality standards, EQSs). Emission values are more easily defined and monitored. EQSs, on the other hand, are more relevant to ecological protection. Clearly, in so far as they can be, emissions from diffuse as compared with point sources, can only be controlled in terms of EQSs. The EQSs are obtained from a procedure akin to predictive risk assessment: using ecotoxicological data, laced with uncertainty factors, to define levels unlikely to have ecological effects. Emission values, on the other hand, are often dictated by what can be achieved technologically.

Restriction and prohibition on production, marketing and use of chemicals as products can be based on risk assessment and cost–benefit analysis. As we shall see, there is certainly an intention to involve risk assessment more and more in this context, and some reference is also made to balancing-based approaches in recent legislation. However, to date most controls in Europe have probably been applied on the basis of "suspicion" and hazard (p. 5).

Controls might also refer to processes. This is not directly related to chemical controls, but it can augment them in a technology-based way (p. 6). Emission values are often set by reference not only to toxicity but also what can be achieved by available technology. The achievement of an EQS is sometimes taken as a minimum requirement to be bettered by available technology.

Controls on packaging, transport, transfer and storage can minimise risk by, for instance, specifying standards for mode of packing and transport, distribution routes and criteria for storage such as security, containment, fire-fighting methodology and so on. Similarly, controls on disposal can specify how the substance and if appropriate its packaging should be disposed of. It can also require other precautions such as labelling, tracking of handlers, mode of container, etc.

So in summary, command and control regulation can be effected through the imposition of standards as:

- Environmental quality targets
- Emission controls
- Product specifications
- Process specifications
- Transport, containment and packaging requirements
- Disposal requirements

These categories are not mutually exclusive: emission controls might be applied to achieve particular EQSs; product specifications might be applied on the basis of disposal characteristics; process specifications might be used to augment emission control.

This regulatory law is concerned with attempting to prevent harm. It is typically enforced by a public agency and can involve a variety of remedies ranging from criminal prosecution to enforcement and variation notices and informal warnings. As usual the punishment is designed to deter. But the whole point of the law is to ensure compliance and protection rather than retribution, so there has been a tradition to try and achieve this through cooperation; informal procedures, rather than confrontation.

3.3   THE INVOLVEMENT OF PRIVATE ACTION

Private action is concerned with remedies for harm done and is privately enforced by the injured party or surrogate agent in their name. There are three main types: compensation for damage done (damages), preventative measures (injunctions) and (rarely) abatement. Rights of action can nevertheless be established through common law or by statutes.

Private law is also often referred to as tort law, a tort being a crooked conduct (a wrong) that causes harm to a specific person. It is extensive and involved and so cannot be dealt with in detail here. However, it has its roots in dealing with disputes over use and abuse of land, and so has obvious implications for environmental protection which require some mention.

The major elements of tort law that have implications for chemical controls are as follows.

3.3.1 Nuisance

There are various forms of nuisance! A **private nuisance** is a tort that involves (1) any wrongful disturbances or interferences with a person's use or enjoyment of land and (2) the act of wrongfully causing or allowing the escape of **deleterious things** on to another person's land. A **public nuisance**, on the other hand, is an act which interferes with the enjoyment of a right to which all members of a community are entitled. Apart from the need to demonstrate an effect over a wide section of the public, there is a broad overlap with private nuisance. One major difference, though, is that a public nuisance in Britain is a criminal offence. **Statutory nuisance** is a system of administrative regulation in which in Britain a local authority is provided with regulatory powers by statute (now through the Environmental Protection Act 1990) and this is backed up by the criminal law. It addresses public issues but these are more precisely defined in the statutes. In the Environmental Protection Act Part III the following categories are relevant to chemical control:

- Smoke emitted from premises so as to be prejudicial to health or a nuisance
- Any dust, steam or smell emitting from an industrial, trade or business premises and being prejudicial to health or a nuisance
- Any accumulation or deposit which is prejudicial to health or a nuisance

The rule in Rylands *v.* Fletcher

This principle arises out of a case reported in 1865 and involves the following:

> ... that the person for his own purposes brings on to his land and collects and keeps there anything likely to do mischief if it escapes, must keep it in at his peril, and if he does not do so, is *prima facie* answerable for all the damage which is the natural consequence of its escape.

This therefore introduces an important principle: that liability is strict, i.e. applies even if there was no intent to do harm. In principle the implications for environmental protection and chemical legislation that follow from this are wide ranging; in practice, however, a number of restrictions placed on the interpretation of the principle in the courts has constrained its effects. One recent development arises out of a decision of the House of Lords in *Cambridge Water Co.* v. *Eastern Counties Leather* in favour of ECL, overturning an earlier decision of the Court of Appeal. The facts were that ECL had operated a tannery in such a way as to allow the escape of significant quantities of organo-chlorine solvent into the groundwater through spillages. This ceased in 1976 but pollution of the water to CWC's borehole was discovered in the 1980s. The House of Lords reasoned that damages in private nuisance or under *Rylands* v. *Fletcher* depend upon *foreseeability* and that this, despite a previous ruling to the contrary by the Court of Appeal, was not a reasonable interpretation here: it was not possible to have foreseen the damages in 1976.

The significance of the ruling is that the highest court in Britain appears to put a barrier in the way of development of strict (i.e. no fault) liability for historic pollution. Nevertheless, strict civil liability for environmental damage embodied in the principle is being considered for application throughout the EU and forms the basis of Superfund legislation in the USA. Indeed the decision from the House of Lords on ECL actually stated that if strict liability is to be developed within this context it is more appropriate for Parliament to carry out this change than the courts.

3.3.2   Negligence

This tort is fault-based, but unlike the other torts considered above does not include the need for a defendant to have a proprietary interest. The plaintiff must establish that a duty of care is owed by the defendant to the plaintiff, that the duty has been breached, and that there was foreseeable damage resulting from the breach.

3.3.3   Trespass

This involves direct interference with personal or proprietary rights without lawful excuse. It has to be intentional or negligent, but there is no need to show damage – all that needs to be shown is interference. So though trespass is often thought of in the context of people entering property without permission, it can, and indeed has, in case law involved substances entering directly into the property of other people. A major requirement, though, is that a causal link has to be established between the directness of the act and the inevitability of its consequences. There are examples of substances released into rivers and the sea being then washed into property where they cause interference. Direct causality has been accepted in the case of rivers since here though discharge and effluent may be separated in space, the unidirectional flow establishes predictable causality. On the other hand, direct causality is less easily established for the marine case where the action of wind, waves and tides makes the ultimate distribution of the effluent less certain.

So, in summary, the main features of tort law that distinguish it from regulatory action are that:

- The right of action is limited in terms of who may bring it and for what reasons – because of its roots it does not seek to protect the environment directly but individual and usually proprietary rights. Environmental protection follows indirectly from the main thrust. The concept of environmental rights *per se* is as yet only poorly developed but is being explored by the EU in proposals concerning Directives on civil liability for environmental damage.
- Controls are normally reactive and compensatory but, through injunctions, can be preventive. Also, the threat of action can act as a potent preventative force.
- The burden of proof is with the plaintiff. In criminal law the prosecuting party (usually the state or its agent) must establish its case beyond reasonable doubt. And in claims of harm caused by pollution this could involve substantial scientific evidence. In tort law plaintiffs are usually required to establish their claim by a preponderance of evidence (balance of probabilities) – as judged by reasonable people. Historically, there has been less call for scientific evidence in this connection.

3.4   ECONOMIC INSTRUMENTS

The Organisation for Economic Cooperation and Development (OECD) has recognised five main kinds of economic instruments – some of which are relevant for chemicals controls.

3.4.1   Charges

There are several possible kinds. Those that recover **administrative** costs, i.e. that relate to the costs of applying the regulations. For example in Britain, the regulatory agencies are required to recover these kinds of costs in establishing and monitoring consents and authorisations in emission controls. There are also charges that relate to the scale of **release** of waste or **use** of environmental resources. For example, pollution charging schemes are operational for water waste discharged to rivers and streams in France, the Netherlands and Germany. The main result is often to raise revenue to cover costs of regulation and to provide subsidies for other action (see section 3.4.3). However, a secondary effect will also be to concentrate the minds of polluters on ways of reducing pollution and hence costs. Finally there are charges that relate to **products**. These are essentially taxes applied to products and services, in a way that reflects the environmental hazards associated with them. The example of the tax differential between leaded and unleaded fuels is now well known. Carbon and energy taxes associated with fuels would be another example.

3.4.2   Creation of markets

The best examples here are tradable quotas used in emission controls. For instance, when there is a ceiling on total emissions, then it is possible to allow allocation by market forces. Trading in SO_2 quotas is now firmly established in the USA as part of the Clean Air Act (see p. 98).

3.4.3   Subsidies

These provide financial assistance for certain pollution control activities – e.g. in the development of more environmentally friendly farming practices. However, subsidies are a departure from the polluter pays principle and can have the effect of disturbing markets. Funding R&D into pollution prevention techniques might be considered as a special form of subsidy.

3.4.4   Deposit/refund schemes

These incentive schemes are often thought of in terms of refunds for returned articles – the containers of goods such as bottles – but another example of more relevance to chemical controls that is currently being researched is a system of environmental assurance bonds. The idea is that an assurance bond equal to the current best estimate of the largest potential future environmental damages would be levied against the producer and kept in an interest-bearing escrow account for a predetermined period. Portions of the bond (plus interest) would be returned if and when the producer could demonstrate that the suspected worst-case damages had not occurred or would be less than anticipated. In the event of damages, the bond would be used in remediation. In this way the cost of a bond would define the price of a right to use the environmental resource in some specified way and would therefore be tradable.

3.4.5   Enforcement incentives

An example would be a non-compliance fee, charged in relation to either profits gained from, or the social cost attributable to, non-compliance. There is little difference from fines.

These market instruments are becoming more fashionable because they provide some flexibility to those regulated in terms of how they achieve requirements, and they send signals into the market-place, so involving the public through consumer choice. They also remove administrative burdens from regulators and should, in principle, reduce the costs of control.

It is, of course, clear that market instruments are based, by definition, on acceptance of some level of pollution, and not all find this acceptable. Moreover, there is the danger that producers might simply decide to incur increased costs for the right to pollute. This depends on setting charges and taxes at a level that will prevent this and lead to the desired level of environmental protection. Targets ought therefore to be based on risk assessment and, where appropriate, cost–benefit analysis. We shall return to this in Chapter 8.

3.5   REGISTERS, "AUDITS" AND LABELS

By influencing consumer choice, information about the environmental credentials of processes or products can act as a potent market instrument for environmental protection. There is also the possibility of direct action, by individuals and lobby groups, on the basis of this kind of information. The information may come in several forms: in public registers, in publicly available statements and on products as labels.

3.5.1   Registers

The US Toxic Release Inventory (TRI) is an example of this kind of instrument. TRI requires industrial sources to report annually their releases to all media. This has elicited risk reduction measures in industry. The disclosure provisions of the TRI are mandatory and accessible to computer-literate citizens. Similar measures are being contemplated in the EU and UK; already the UK Environment Agency publishes an annual Chemical Release Inventory which is a computerised database of industrial pollution arising from major industrial plants regulated under integrated pollution control legislation.

3.5.2   Audits

There is an increasing trend towards encouragement of business to introduce and implement environmental management systems (e.g. by the EU, p. 87). These usually require statements about performance, both in terms of compliance with legislation *and* in terms of environmental effects. The British Standards Institute Environmental Management Standard BS 7750, for example, requires the establishment of both a compliance and an environmental *effects* register. The procedure for assessing effects can include consideration of:

- Controlled and uncontrolled emissions to atmosphere and discharges to water
- Wastes

- Contamination of land
- Use of natural resources
- Effects on specific parts of the environment and ecosystems

This is intended to include effects arising from normal and abnormal operating procedures; past, current and future activities. There are clear chemical aspects to many of these measures, and an obvious relationship with the inventories mentioned above.

The purist will note, however, that most on the above list are not actual environmental effects; rather they are causes of effects. To translate emissions, waste production and land contamination into measures of effects will require either hazard identification or risk assessment. There may also be a requirement to combine all effects measures from an operation to get some impression of overall, potential impact and overall changes in performance through time.

Just exactly how much of the detail of these registers is made available to the public may well depend upon the extent to which industry feels protective about commercially sensitive information. Minimal achievement of the requirements of standards such as BS 7750 will be signalled by the appropriate labels (see below).

3.5.3   Labels

There are two kinds of labels: those that signal harm or the potential for causing harm to the environment, and those that suggest that the products labelled are more environmentally friendly than are functionally equivalent products.

Negative warnings of the former kind are usually based on hazard identification. The logic here is that labels should indicate potential to cause harm – the hope being that this influences use to an extent that reduces the likelihood of that potential being realised, i.e. hazard information leads to reductions in risk.

Positive labels – often referred to as ecolabels – are assigned with more difficulty and have to take into account potential effects from the product throughout its life-cycle, from raw materials through production and product use to disposal. These life-cycle assessments, as they are known, therefore potentially *involve a considerable amount of work* and hence are *potentially expensive*. Minimally they involve constructing an inventory of energy used and waste produced at each stage of production and use. There ought also to be an estimate of environmental impact that should involve hazard identifications and, possibly, risk assessments, and as in "auditing" all measures for all life-cycle components should ideally be boiled down into a single number for comparative purposes. This is still an active area of research. Clearly, those products with *least impact per unit function* (e.g. per unit cleaning power of a detergent) are the ones that will receive positive labels.

3.6   VOLUNTARY AGREEMENTS

Voluntary action can take one of at least two forms: concerning a decision on whether to get involved in a programme or concerning decisions on how to achieve required targets.

Examples of the former are decisions on whether to be involved in the ecoaudit and ecolabel schemes mentioned above. These are voluntary schemes, but once a decision has been made to be involved the requirements are strict and open to independent checking and public scrutiny. A less formal voluntary scheme of a similar kind is the Chemical Industries Association (CIA) Responsible Care Programme. Companies that sign up to this commit themselves to a number of principles (Box 3.1) that are subject to self-monitoring and reporting.

Voluntary action of the other form, of deciding on ways of achieving specified targets, is an approach that is adopted by the Dutch in the development of their National Environmental Policy Plan, initiated in 1989. Here the government holds talks with involved parties to formulate policies. The objectives, for example of environmental quality standards, are laid down by government; it is the method and timescale of achieving these that are open for discussion. These are agreed formally and written into covenants with performance indicators that are monitored accordingly. Methods of achieving targets could obviously involve the provision of

Box 3.1   CIA Responsible Care Programme

The following Guiding Principles form the basis of this commitment:

- Companies should ensure that their health, safety and environment policy reflects the commitment and is clearly seen to be an integral part of the overall business policy.
- Companies should ensure that management, employees at all levels and those in contractual relationships with the Company are aware of the commitment and are involved in the achievement of their policy objectives.
- All Company activities and operations must be conducted in accordance with relevant statutory obligations. In addition, Companies should operate to the best practices of the industry and in accordance with Government and Association guidance.

In particular, Companies should:

- Assess the actual and potential impact of their activities and products on the health and safety of employees, customers, the public and environment.
- Where appropriate, work closely with public and statutory bodies in the development and implementation of measures designed to achieve an acceptably high level of health, safety and environmental protection.
- Make available to employees, customers, the public and statutory bodies, relevant information about activities that affect health, safety and the environment.

Members of the Association recognise that these Principles and activities should continue to be kept under regular review.

information (labels), training, methods of production, packaging, etc. Another example of this is a voluntary agreement, hatched in 1995, between industry and the OECD within the context of the latter's risk reduction programme (see p. 110) with respect to restrictions on the manufacture of selected brominated flame retardants. Industry agreed to stop manufacturing selected substances and committed itself to employing best available techniques not entailing excessive costs for improving purity, minimising losses and exposure. Furthermore, industry agreed to operate on the basis of the Responsible Care Programme (above) and especially to emphasise product stewardship. There was a commitment to monitor compliance and effectiveness by industry (with reports being submitted to the OECD on a regular basis) but also the option of member countries to carry out their own monitoring. This is the first time that a voluntary agreement has been reached on specific substances, at least in an international context.

An advantage of this voluntary approach is that industry can identify cost-effective ways of achieving given targets. Moreover they may enhance the speed and flexibility with which controls can be applied. Regulations can take years to negotiate and bring into force (as will be apparent later), and they may be rapidly outdated by developments in scientific understanding and in technological developments and management techniques. On the other hand voluntary agreements can be implemented more rapidly and hence be more responsive to changing conditions. They may require some background legislation to ensure compliance by all participating companies since cheating might bring short-term economic advantage. For the sake of credibility they will also require acceptable and transparent monitoring and reporting systems.

3.7 RÉSUMÉ

- Risks arise from chemical substances at all points along the supply chain from production to use and ultimate disposal.
- Releases might occur from normal or abnormal operations and uses and also from accidents.
- Releases might be from point sources or from more distributed sources.
- Methods of control arise from specific legislation through market instruments of various kinds to voluntary agreements.
- Table 3.1 summarises these features and indicates where the emphases might be in the various components.

3.8 FURTHER READING

Ball, S. & Bell, S. (1994). *Environmental Law*, 2nd edn. Blackstone Press Ltd, London.
Cranor, C. F. (1993). *Regulating Toxic Substances*. Oxford University Press, Oxford.
McGregor, G. I. (1994). *Environmental Law and Enforcement*. Lewis Publishers, Boca Raton.

4

European and UK axis

4.1 EC, EU AND ENVIRONMENT

In the European Community (EC), now Union (EU), the legislation of Member States is importantly influenced by the common legislation of the group. This chapter briefly describes the basis of the European legislation and how it interacts with Member State legislation as an important preliminary to considering UK and EU chemical controls legislation.

The Treaty of Rome (EEC Treaty; established the European Economic Community (EEC) in 1957) was about a common trading market. However, harmonisation of national laws to prevent trade barriers (Article 100 EEC) was used as a basis for justifying laws relating to pollution control. Article 235 (EEC) relating to general and residual powers was also used as a justification for environmental protection (e.g. a Directive on wild birds was legitimised on this basis in 1979). The Single European Act (SEA; came into force 1986) amended the EEC Treaty by introducing explicit environmental law-making powers in Articles 130r, 130s and 130t. It also introduced Article 100a which requires that the issuing of legislation for the approximation of measures within Member States with respect to the functioning of the internal market will take as a base a high level of protection concerning the environment (paragraph 3). The Treaty on European Union (TU) extends environmental responsibility further and introduces some modifications to decision-making procedures that will be considered further below. It was signed in Maastricht in February 1992 but not formally ratified by all Member States and adopted until November 1993 and consists of two segments. The first – on common provisions – stands alone. The second amends the EEC Treaty in detail and renames it the Treaty Establishing the European Community (EC Treaty). In the second segment it introduces a modification to Article 2 (EC Treaty) that states that the Community will promote "a harmonious and balanced development of economic activities, sustainable and non-inflationary growth respecting the environment".

The TU created the European Union. This rests on three pillars: common foreign and security policy; home affairs and justice policy; all policies previously carried out under the terms of the EEC Treaty. The first two pillars are new. The last one is as redefined in the amended EC Treaty. It remains correct therefore to refer to the latter as EC policy. At the same time, since the EC is part of the European Union, it also remains correct to refer to the EU. The Treaty of Rome, on the other hand, initially created the European Economic Community and this is explicitly replaced in the TU by European Community, so EEC is no longer appropriate. One last complication: the European Commission, an institution of the EC/EU (see below), is also sometimes abbreviated to EC, but more often to CEC.

At the time of writing the Intergovernmental Conference (IGC) of Member States is examining the provisions of the Treaty and this may lead to further modifications in policies and procedures over the next few years.

4.2   THE INSTITUTIONS OF THE EU

The four main institutions are:

1.   European Commission – executive of the EU – **it proposes policy**. It consists of a College of Commissioners including the President, that are appointed by Member States but allocated portfolios by the President. The President is appointed by the Council of Ministers – subject to ratification by Parliament as is the College of Commissioners. The bureaucracy is divided into a number of Directorate-Generals, each of which is headed by a Director-General. Directorate-General (DG) XI deals with environment, consumer protection and nuclear safety.
2.   Council of Ministers – political body made up of representatives from each Member State – **it adopts legislation**.
3.   European Parliament – has been mainly **consultative and advisory** but does have **powers over legislation** which were increased under the TU.
4.   European Court of Justice – collection of judges appointed by common agreement of Member States – **has supreme authority on matters of EC law**.

Another institution of rising importance is:

•   The Court of Auditors – consists of one member from each Member State. Its task is to examine all accounts within the EU framework and also has the responsibility of ensuring sound financial management.

Also relevant for chemical controls are the European Environment Agency and the European Chemicals Bureau. The former was established by Regulation (EEC) No. 1210/90 in 1990 and is charged with the collection and collation of data to describe the quality of the environment, pressures on the environment and the sensitivity of the environment. It is located in Copenhagen. The Bureau, located within the Joint

Research Centre, Environment Institute at Ispra (Italy), provides scientific and technical support for the Commission (in particular DG XI) associated with information gathering on chemicals subject to new and existing substances legislation to be described below.

4.3   INVOLVEMENT OF INSTITUTIONS IN LEGISLATION

In a nutshell: the Commission proposes policies and legislation whereas the Council decides on policies and legislation in terms of adopting or rejecting them. The European Parliament gives its opinions on legislative proposals – but prior to 1986 the Council could take or leave these. Nevertheless, about a quarter of parliamentary amendments did find their way into EC law. The Council may specifically authorise the Commission to implement legislation through subsidiary legislation. This is usually used to amend technical annexes to the original legislation. Member States participate in the process through one of a series of procedures laid down by Council (Decision 87/373/EEC).

The Single European Act introduced a **cooperation procedure** which gave Parliament more powers, and the new Treaty on European Union introduces a **co-decision procedure** that confers even more powers on the Parliament.

So, under the EC Treaty, there are now three procedures for adopting legislation that differ in the form of voting in the Council and on the extent to which parliamentary views are taken into account:

(A)   **Unanimous voting** by Council – the Parliament simply gives an opinion. This covers only a limited number of measures:
- Fiscal
- Town and country planning, land use (not waste management)
- Those significantly affecting choice between energy sources and the general structure of the energy supply within the Member States

(B)   **Cooperation procedure** applies to:
- Most environmental legislation under Articles 130r, s and t (except those specified under (A))
- Other aspects of environmental policy where the Council agrees unanimously to use the majority voting procedures
- Individual R&D programmes

The procedure (summarised in Figure 4.1) allows the Parliament to give an opinion on the Commission's proposals **and** the Council's amendments (as encapsulated in its Common Position) before the Council has reached its final decision. So the Parliament can send signals to the Council and the Commission to rethink. Under this procedure the Council operates by Qualified Majority Voting (i.e. a complex system in which votes are weighted according to population size of Member States).

(C) **Co-decision procedure,** applies to:
  • Framework R&D programmes
  • Large trans-European infrastructure projects
  • Public health
  • General action programmes setting out priority objectives (see p. 52)
  • Internal market legislation under Article 100a

It is illustrated in Figure 4.2. It involves two or possibly three parliamentary readings and allows Parliament to reject (not to accept) legislation. Qualified majority
voting is again involved in Council decisions.

So to summarise, the major routes for environmental legislation through the EC
Treaty are as follows.

Article 100a

  • Article 100 enables legislation to be made that seeks harmonisation, i.e. functioning of a common market.

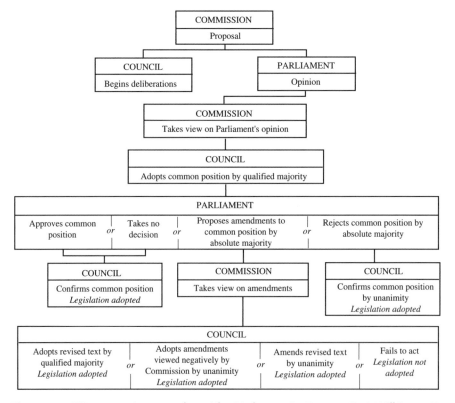

Figure 4.1   EC cooperation procedure. After Verhoeve, B., Bennet, G. & Wilkinson, D.
(1992). *Maastricht and the Environment.* Institute for European Environmental Policy, p. 44

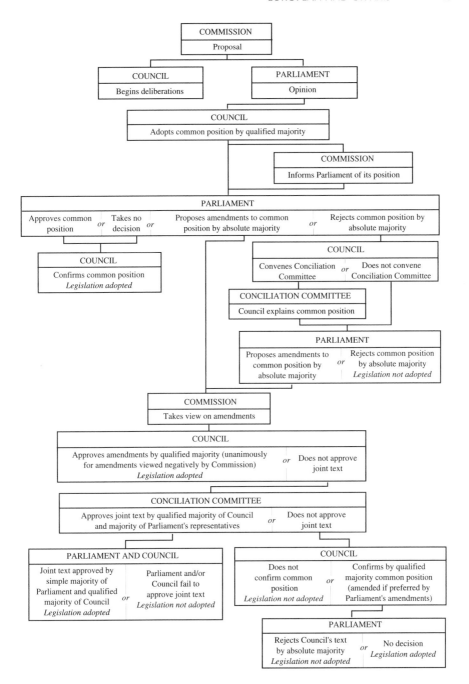

Figure 4.2 EC co-decision procedure. After Verhoeve *et al.*, p. 45. See Figure 4.1

- Article 100a introduced qualified majority voting for this purpose – TU moves 100a to "co-decision" (with its parliamentary veto).
- In general Article 100a requires uniformity throughout Member States in approving legislation. There is one exception (100a (4)); national deviation if justified on the grounds of major needs or relating to the protection of the environment and the Commission verifies this. This has been invoked by Germany in justifying regulations on pentachlorophenol (PCP). In 1989 it adopted regulations banning the sale and use of PCP without exception. The Directive on PCP (p. 81) allowed some exceptions and Germany sought derogation under paragraph 4. The Commission agreed, but the Court of Justice, to whom the issue had been directed by the French, did not; however not because they necessarily judged that the use of paragraph 4 was unsound but because the Commission had not given enough detail to justify its position. Accordingly the Commission came back with a more detailed argument on why the Germans were right to maintain tougher laws for Germany – including evidence of elevated historical exposure here. More recently Denmark has also invoked paragraph 4 to maintain an effective total ban on PCP.

Articles 130r, 130s, 130t

- These provide specific justification for environmental protection laws, even where there is no direct link to economic aims.
- They were previously adopted by unanimous voting but under TU largely by majority voting and two parliamentary readings (the cooperation procedure).
- Article 130t specifically provides stricter measures than those agreed under Article 130s may be employed by Member States – but the Commission shall be informed.

In other words Article 100a seeks harmonisation and equality; whilst Articles 130r, s, and t seek minimum or threshold standards.

Other general provisions that may be of relevance for environmental protection are **Article 213**, giving the Commission powers to collect information to carry out any checks required for the performance of its tasks, and **Article 235** providing "necessary powers" to take "appropriate measures" to obtain objectives of the Community and the Treaty. The former is subject to limits "laid down by the Council". The latter is subject to consultation with the European Parliament.

    Article 36 is also worthy of note. This allows prohibitions or restrictions on imports and exports of goods in transit on grounds of public morals, public policy or public security, the protection of health and life of humans, **animals or plants**, etc., provided such prohibitions or restrictions do not constitute a means of arbitrary discrimination or a disguised restriction. In the 1980s the Commission challenged the Danish law that required beer and soft drink containers to be returnable – arguing that this established a hidden trade barrier against foreign manufacturers.

But the Court of Justice found in favour of Denmark on the basis of Article 36. The Court was of the view that protection of the environment is one of the mandatory requirements of the EU.

4.4   THE LEGAL INSTRUMENTS

There are several kinds of instruments. What form they take determines how they are implemented. They are as follows:

- Regulations – legislative acts of general application – normally directly effective without the need for national facilitating legislation. These enter into force on the date specified in them or failing that on the twentieth day following the publication in the *Official Journal* (*OJ*).
- Directives – are addressed to Member States and are binding as to the results to be achieved – but choice on how to implement them is left to Member States, through their own domestic law. Normally implementation is required within a time limit.
- Decisions – are binding on the group to whom they are addressed and may also be directly effective.

There are also recommendations, resolutions and opinions. However, these are not binding. The instruments are prefixed by "Council" or "Parliament and Council" (when done under co-decision) or "Commission" to signal their source. All instruments are published within the *OJ* of the EU. This is published in all languages of Member States and is subject to a formal mechanism of translations prior to publication. It is made up of three series, C, S and L, and it is the L series that carries the legislation.

The nomenclature for Regulations is of the following general pattern: (EEC) No. 793/93 which indicates that it was adopted under the EEC Treaty (would now be EC) and was the 793rd piece of legislation recorded in the *OJ* for 1993 when it was adopted. The nomenclature for directives is the other way round: 67/548/EEC.

All Member States are legally bound under the Treaties to adopt binding legislation within the prescribed time limits. However, they do this with varying levels of success and there has been some attempt to create cooperation between regulatory authorities within Member States to monitor and improve implementation in the environmental context.

If a Member State does not implement legislation or does not implement it properly it is for the Commission to start infringement proceedings. These are phased. First the Commission writes informally drawing attention to the infringement. Failing subsequent action it will issue a reasoned opinion explaining what it believes are the main features of non-compliance. Failing compliance the Commission can bring a Member State before the Court of Justice. This is the sole arbiter of the issue and following the TU has the ultimate power to impose fines for non-compliance. To date this has not been used, but it is being contemplated by the Commission in a number of cases.

4.5   EVOLUTION OF ENVIRONMENTAL POLICY

The Commission sets out its philosophy for policy frameworks in the form of wide-ranging action programmes. There have been five to date. The Council issues resolutions on these, and some of the principles have been assimilated into the Treaties by the SEA and TU.

The first and second action programmes, adopted (by the Commission) respectively in 1973 and 1977, provide lists of actions to be taken to control a wide range of pollution problems. They list 11 major principles:

1. Prevention is better than cure.
2. Environmental impacts should be taken into account at the earliest possible stage in decision making.
3. Exploitation of nature which causes significant damage to the ecological balance must be avoided.
4. Scientific knowledge should be improved to enable action to be taken.
5. The polluter should pay.
6. Activities in one Member State should not cause deterioration of the environment of another.
7. Environmental policy in the Member States must take into account the interests of the developing countries.
8. The EC (now EU) and its Member States should promote international and world-wide environmental protection through international organisations.
9. Environmental protection is everyone's responsibility, therefore education is necessary.
10. Environmental protection measures should be taken at the most appropriate level – the subsidiarity principle.
11. National environmental programmes should be co-ordinated on the basis of a common long-term concept and national policies should be harmonised within the Community not in isolation.

The third action programme, adopted in 1983, attempted to provide an overall strategy for environmental protection, shifting the emphasis from control to prevention. The fourth action programme (adopted in 1987) emphasised:

- Effective implementation of existing legislation
- Regulation of all environmental impacts of "substances" and "sources" of pollution
- Increased public access to and dissemination of information
- Job creation

The latest, fifth, action programme was eventually made public in 1992. The title – **Towards Sustainability** – sets the tone, recognising that though previous programmes have made progress in environmental protection there is still work to be done; as there will be even after the new programme has run its course (time frame to 2000).

There are two important departures from previous programmes, that are of importance:

- First – there is an explicit recognition of the importance of involving all sectors of society in environmental protection in a "spirit of shared responsibility". Less command and control is advocated and more *persuasion* by other instruments.
- Second – previous programmes have concentrated on applying controls environmental sector by sector. This new programme concentrates instead on economic sectors – in particular industry, energy, transport, agriculture and tourism. This recognises the need to take account of simultaneous impacts to all parts of the environment, in an integrated way, when attempting to regulate activities.

As part of a "mid-term" review in January 1996 the Commission proposed five key priorities to further the aspirations of the programme: improve integration of environment and other policies especially in agriculture, transport, energy and industry; broaden the range of instruments with a proposal for framework legislation establishing common principles for voluntary programmes; improve implementation and enforcement of environmental instruments; raise general awareness of sustainable development issues; reinforce the EU's role in international forums.

Amendments to the EEC Treaty by the SEA incorporated some of the general principles from the Action Programmes within Articles 130r, s and t. These goals and elements of environmental protection action are:

1. To preserve, protect and improve the quality of the environment.
2. To contribute towards protecting human health.
3. To ensure a prudent and rational utilisation of natural resources.

EC environment protection actions must become integrated into other policies based on three principles:

4. Preventive action should be taken.
5. Environmental damage should be rectified at source.
6. The polluter should pay.

To these general principles the TU adds an international dimension to the list of objectives:

- Promoting measures at an international level to deal with regional or worldwide environmental problems and develops policy that **shall**

- Aim at a high level of protection . . . and . . . **shall be based** on the precautionary principle

So, in summary, the SEA introduced the principles of "prevention rather than cure" and "polluter pays"; TU introduced the "precautionary principle". At the same time (under paragraph 3 of Article 130r) it is stated that in preparing its policy relating to environment the Community **shall take account** of:

- Available scientific and technical data
- Environmental conditions in the various regions of the community
- The potential benefits and costs of action or lack of action
- The economic and social development of the community as a whole and the balanced development of its regions

The phrases "shall be based on" and "shall take account of" (above) are as used in the Treaty and imply that precaution takes precedence over both scientific and economic considerations in developing environmental policy. This potentially creates some confusion in the development of environmental policy (see p. 6). There may be changes to these and other aspects of the Treaty relating to environment and sustainability following the IGC.

4.6   RÉSUMÉ

- Though established to ensure a free trade market the European Community, now Union, quickly became involved in developing standard environmental protection measures across Member States.
- These are based on a number of important principles: e.g. that prevention is better than cure; that the polluter should pay; that precaution should be exercised; but that scientific understanding and economic welfare should also be taken into account.
- Principles, policy and legislation are proposed by the Commission, but adopted by Council with important involvement of the Parliament.
- Powers for adoption of measures come largely through two routes: Article 100a concerned primarily with preventing the development of trade barriers and 130r concerned with environmental protection *per se*.
- Measures emerge as regulations that apply directly to Member States or directives that have to be applied through national legislation according to a defined time-table.
- Some modifications of the above may emerge over the next few years as a result of the IGC.

4.7 FURTHER READING

Ball, S. and Bell, S. (1994). *Environmental Law*, 2nd edn. Blackstone Press Ltd, London.

Commission of the European Communities, Directorate-General XI (1992). *European Community Environmental Legislation*. Official Pubs. European Communities, Brussels, Luxembourg (one of 7 volumes).

Haigh, N. (1995). *Manual of Environmental Policy: the EC and Britain*. Cartermill, London. (Regular updates.)

UK Government (1996). *A Partnership of Nations. The British Approach to the European Union Intergovernmental Conference 1996*. HMSO, London.

5

Specific legislation in Britain and Europe

5.1 INTRODUCTION

In moving from general principles to particular legislation there is always a great danger that subsequent changes might render the information out of date. This is particularly so with environmental legislation where the pace of change is rapid, so this chapter should be treated as a guide to where the legislation was at the time of writing. It is therefore not necessarily the last word on the subject.

In both the EU and UK it is first of all important to note that the key pieces of legislation are of the framework kind, i.e. laying down the general rules but requiring more specific instruments of legislation to address particular problems. As far as the EU is concerned, the specifics come from amending legislation and daughter directives and in implementing national legislation. In the UK, specifics come from statutory instruments. In both cases, important interpretations can also come from less formal documents, such as guidance notes. The statutory instruments have to be prepared according to specifications defined in the appropriate Acts. They are laid before Parliament, but not usually debated. They are generally called Regulations – sometimes Orders – and the former are a potential source of confusion with EU instruments.

For Britain, the Environmental Protection Act 1990 (EPA 90) is a good example of framework legislation, addressing a wide range of areas. Not all of it became effective when it was passed in 1990; parts required special statutory instruments, commencement orders, to initiate them. Thus elements of Part II of the Act which deal with waste management were only implemented in 1994, so earlier legislation embodied in the Control of Pollution Act (COPA), which EPA 90 was intended to replace, remained in force until that time. Section 143 (Part VIII), requiring the establishment of a system of registers of potentially contaminated land, was never implemented because of a fear of land blight as a consequence of it and has been superseded by provisions in the Environment Act 1995. Phased implementation of a wide-ranging Act such as EPA 90 is important to allow time for both regulators and regulated to adjust, but it also allows government to manipulate implementation of legislation that has otherwise received parliamentary approval.

The Water Resources Act 1991 and Water Industries Act 1991 (superseding the Water Act 1989) also represent wide-ranging legislation concerning the management and protection of water resources and supplies, establishing the National Rivers Authority (NRA) as regulatory body. The Environment Act 1995 though initially conceived as legislation to establish regulatory environment agencies (assimilating existing bodies such as the NRA and HMIP – see Glossary), in the end addressed a number of other issues including contaminated land, national parks, air quality and a range of miscellaneous items.

Some other pieces of UK legislation that are involved with chemical controls and that have an environmental aspect are:

- Health and Safety at Work, etc. Act 1974 – concerned largely, but not exclusively, with the working environment
- Consumer Protection Act 1987 – concerned with product safety
- Food and Environment Protection Act 1985 – concerned with food standards, but also importantly with controls on pesticides

Note also should be made of the European Communities Act 1972 in which section 2(1) provides that EC legislation is recognised as law in Britain, and section 2(2) enables EC legislation to be transposed into domestic law, even if there is no present statute authorising it.

We now consider the specific legislation that is concerned with protecting the environment from chemical pollution. The emphasis will be on industrial chemicals (section 1.1), but for completeness reference will also be made to agrochemicals, detergents, oil products and solid wastes that contain hazardous chemicals. Following the logic discussed in Chapter 1, we distinguish between information-gathering and controlling instruments, and within the latter between point-source (pollution) and distributed control. In this we shall begin with a description of the EC legislation and consider how it has been translated into national legislation. Later in the chapter we shall consider if and how all the separate pieces of legislation fit together. The controls turn out to be somewhat piecemeal, referring to particular substances in particular media. In a final section there is consideration of trends towards more holistic and integrated controls and these are taken up again in Chapter 8.

5.2   INFORMATION GATHERING AND PRESENTATION AS LABELS

As far as potentially hazardous chemicals go in general, there are two pieces of important information-gathering legislation: one, a Directive, that is concerned with labelling chemicals and providing a mechanism for obtaining necessary information on so-called new chemicals, and the other, a Regulation, concerned with obtaining information on so-called existing chemicals. There is also some specific legislation for pesticides.

5.2.1   Labelling, and new chemicals

Directive 67/548/EEC amended for the seventh time by 92/32/EEC on the approximation of the laws, regulations and administrative provisions relating to the classification, packaging and labelling of dangerous substances.

Legal basis: initially Article 100 and now 100a.

Aims

As its legal basis (100/100a) suggests, the main aim of this Directive is in avoiding disputes between Member States in classifying, labelling and packaging dangerous substances and notifying new ones, that might lead to trading barriers and unequal competition. Before being marketed all dangerous substances must be classified and labelled on the basis of their physico-chemical and toxicological properties. The original Directive was concerned only with those substances posing risks to human health – in terms of human toxicity and special hazards such as explosives – specified on a list in its Annex 1. The Sixth Amendment (79/831/EEC) required labelling of all substances, including new ones, that would be done on the basis of information notified by the manufacturer or supplier. This therefore necessitated the specification of a notification scheme and the definition of what constitutes new chemicals. The latter was achieved by the compilation of all known commercial chemicals to September 1981; known as the European Inventory of Existing Commercial Chemical Substances (EINECS; published in *OJ*, C **146A**, 1990; notice that this is in the C series of *OJ* since it is a notice, not a piece of legislation). Chemicals not on this list are designated as new chemicals. Chemicals so notified are summarised on another list: the European List of Notified New Chemical Substances (ELINCS – first published in *OJ*, C **130**, 1993 and to be updated annually).

The Sixth Amendment introduced a classification category "dangerous for the environment" that was explicitly defined much later in a so-called adaptation (the twelfth) to technical progress (see below) of the Directive (91/325/EEC). A Seventh Amendment (92/32/EEC) specified a label for these substances. It also placed a duty on competent authorities to carry out risk assessments on substances notified. This is done according to methodology defined in another Directive (93/67/EEC).

Amendments to the parent Directive are made by the Council according to the principles laid down under Article 100a. Adaptations, including the formal listing of substances on Annex 1, are carried out by committee according to procedures defined within the parent Directive.

Details

The evolution of the parent Directive has been somewhat tortuous with relevant amendments and daughter directives specifying adaptations scattered through the *OJ*s. From an environmental point of view, it is probably better to begin by examining its approach to the notification of new substances even though, as already noted, this was developed in support of the primary concern with classification and labelling.

Notifications

Those wishing to market a new substance (at greater than 1 tonne p.a.) have to submit to the competent authority: a technical dossier supplying information necessary for evaluating the foreseeable risks (see below) for man and the environment – the level of detail and the extent of testing required depend upon amounts likely to be released on to the market (Box 5.1); a declaration concerning any unfavourable effects in terms of foreseeable uses; the proposed classifications and labelling (see below); in the case of dangerous substances a proposal for a safety data sheet "to enable professional users in particular to take necessary measures as regards protection of the environment etc.". For substances marketed at <1 tonne but >100 kg per manufacturer/year, a reduced form of the technical dossier is allowed with minimum ecotoxicological information – but still the other items listed above are required. For substances marketed at < 100 kg per manufacturer/year the notifier can restrict the information in the technical dossier further and this then requires no ecotoxicological data. For substances marketed at >1 tonne per manufacturer/year, they can be put on the market 60 days following submission in the absence of any indications to the contrary. For substances marketed at <1 tonne per manufacturer/year the time limit is 30 days.

An indication of the kinds of ecotoxicity tests required under this legislation has already been given in Box 2.1. Council Directive 87/18/EEC requires that laboratories carrying out tests should do this in accordance with good laboratory practice (GLP) originally laid down by the OECD (see p. 109). The essence of GLP is that all experimental procedures are carried out such that results can be verified and in principle, if necessary, repeated. Laboratories are required to certificate that their work is carried out according to GLP and Member States have to adopt measures necessary for verification. Laboratory inspections and certifications for GLP in the UK are carried out by the Department of Health.

Classification

The Directive 92/32 lists 15 categories of danger, that includes dangerous for the environment, "i.e. substances which, were they to enter the environment would present an immediate or delayed danger for one or more of the components of the environment".

Box 5.1   Some of the requirements for notification under 67/548 legislation

Levels (per year per manufacturer)	Annex (defining information required in notification)	Requirements	Annex (defining tests)
< 100 kg	VIIC (base set)	Name	
		Molecular and structural formula	
		Composition	
		Methods of detection / determination	
		Production	
		Proposed uses	
		Estimated production and/or imports	
		Recommended methods + precautions in handling, etc.	
		Emergency measures	
		Physico-chemical properties	VA
		+ some toxicity studies	VB
< 1 tonne	VIIB (base set)	As VIIC + some more physico-chemical information, more elaborate tox + mutagenicity	VA/VB
		+ ECOTOX (biotic deg)...	VC
1–10 tonnes	VIIA (base set)	As VIIB + some more physico-chemical information	VA/VB
		More elaborate tox...	
		+ mutagenicity	
		+ ECOTOX	VC
		Fish	
		Daphnia	
		algae	
		Bact inhibit: when biodeg may be affected by inhib of bact.	
		Deg...	
		Absorption/desorption	

> 10 tonnes: Some or all of Level 1 that involves some more long-term ecotoxicity tests (specified in Annex VIII).
> 1000 tonnes: A programme of test studies, Level 2, agreed with competent authorities (guided by Annex VIII).

The level of danger is judged by reference to "intrinsic properties" (Article 4(i)) which, in the case of new substances are assessed in terms of the results from the ecotoxicological tests, i.e. the classification is hazard based but, of course, the thoroughness of the hazard assessment depends, as indicated in Box 5.1, on amounts placed on the market and hence on an index of exposure. Moreover, biodegradability is also taken into account in the classification and this will also influence exposure.

Figure 5.1   EC label signifying dangerous to the environment

Labelling

Every package containing dangerous substances must show:

- The name of the substance
- Name and address, etc. of those marketing the substance
- Danger symbols (Figure 5.1)
- Standard risk (R) phrases, the form of which depends upon the results of ecotoxicological tests (Box 5.2)
- Standard phrases (S-phrases) relating to safe use (Box 5.3)

The Directive also requires that all substances which are classified as dangerous should, at the time of first delivery to consumers, be accompanied by a safety data sheet that contains such information as is necessary to protect them and the environment. Detailed rules for this are set out in 93/112/EEC.

In designating a label with phrases it will be clear from the foregoing that the Directive recognises three kinds of substances:

- Those on Annex 1 – when R- and S-phrases are predetermined (under review)
- Those on EINECS, but not on Annex 1 (i.e. existing substances) when a label is assigned by those marketing on the basis of the principles laid down in the Directive but using only existing evidence (known as self-classification)
- Those not on Annex 1 or EINECS (i.e. new substances) when a label can be assigned on the basis of the information in the technical dossier

As regards substances appearing in Annex 1, the label should also include the words "E(E)C label", otherwise labels are provisional. Substances provisionally labelled can be incorporated into Annex 1, by technical adaptation procedures. The authority receiving the notification for a dangerous substance is responsible for submitting a formal proposal for the entry to be included in Annex 1. This proposal is communicated by the Commission to the other Member States which have six months to send comments to the originating authority. That authority then has a further three months to reconcile the various comments and to prepare a final proposal. This is then submitted by the Commission for approval to the committee for adaptation to technical progress. Once approved the substance is introduced to Annex 1 and thereafter it is illegal to classify and label substances in any way other than specified. Finally, the sizes and colours of labels are specified in the Directive.

Box 5.2   Definition of standard risk (R) phrases under 67/548 legislation[1]

For the purpose of classification and labelling, substances are divided into two groups according to their acute and/or long-term effects.

Aquatic systems

Substances should be classified as dangerous for the environment, assigned the corresponding symbol (Figure 5.1) with the indication of danger "dangerous for the environment" and the appropriate risk phrases in accordance with the following criteria:

R50:	Very toxic to aquatic organisms and
R53:	May cause long-term adverse effects in the aquatic environment

Acute toxicity:	96 hr LC_{50} (for fish):	$\leqslant 1$ mg/l
	or 48 hr EC_{50} (for *Daphnia*):	$\leqslant 1$ mg/l
	or 72 hr IC_{50} (for algae):	$\leqslant 1$ mg/l
and	the substance is not readily degradable or the log P_{ow} (log octanol/water partition coefficient) (unless the experimentally determined BCF	$\geqslant 3.0$ $\leqslant 100$)

R50:	Very toxic to aquatic organisms

Acute toxicity:	96 hr LC_{50} (for fish)	$\leqslant 1$ mg/l
	or 48 hr EC_{50} (for *Daphnia*):	$\leqslant 1$ mg/l
	or 72hr IC_{50} (for algae):	$\leqslant 1$ mg/l

R51:	Toxic to aquatic organisms and
R53:	May cause long-term adverse effects in the aquatic environment

Acute toxicity:	96 hr LC_{50} (for fish):	1 mg/l $< LC_{50} \leqslant 10$ mg/l
	or 48 hr EC_{50} (for *Daphnia*):	1 mg/l $< EC_{50} \leqslant 10$ mg/l
	or 72 hr IC_{50} (for algae):	1 mg/l $< IC_{50} \leqslant 10$ mg/l
and	the substance is not readily degradable or the low P_{ow} (unless the experimentally determined BCF $\leqslant 100$)	$\geqslant 3.0$

Substances should be classified as dangerous for the environment and assigned the appropriate risk phrases in accordance with the following criteria:

R52:	Harmful to aquatic organisms and
R53:	May cause long-term adverse effects in the aquatic environment

Acute toxicity:	96 hr LC_{50} (for fish):	10 mg/l $< LC_{50} \leqslant 100$mg/l
	or 48 hr EC_{50} (for *Daphnia*):	10 mg/l $< EC_{50} \leqslant 100$mg/l
	or 72 hr IC_{50} (for algae):	10 mg/l $< IC_{50} \leqslant 100$mg/l
and	the substance is not readily degradable. These criteria apply unless there is sufficient additional scientific evidence concerning degradation and/or toxicity to provide adequate assurance that neither the substance nor its degradation products will constitute a potential long-term and/or delayed danger to the aquatic environment.	

(*continued*)

Box 5.2   (*Continued*)

> Such additional scientific evidence should normally be based on the studies required at Level 1 or studies of equivalent value, and could include:
>
> (i)   a proven potential to degrade rapidly in the aquatic environment;
> (ii)   an absence of chronic toxicity effects at a concentration of 1.0 mg/l, e.g. a no-observed effect concentration of greater than 1.0 mg/l determined in a prolonged toxicity study with fish or *Daphnia*.
>
> Substances not falling under the criteria defined above but which on the basis of the available evidence concerning their toxicity, persistence, potential to accumulate and predicted, or observed, environmental fate and behaviour may nevertheless present an immediate or long-term and/or delayed danger to the structure and/or functioning of aquatic ecosystems should be assigned at least one of the following phrases:
>
> R52:   Harmful to aquatic organisms
> R53:   May cause long-term adverse effects in the aquatic environment
>
> Poorly water soluble substances, i.e. substances with a solubility of less than 1 mg/l, will be covered by these criteria if:
>
> (a)   they are not readily degradable and
> (b)   the $\log P_{ow} \geq 3.0$ (unless the experimentally determined BCF ≤ 100)
>
> The criteria in (a) and (b) above apply unless there is sufficient additional scientific evidence concerning degradation and/or toxicity sufficient to provide adequate assurance that neither the substance nor its degradation products will constitute a potential long-term and/or delayed danger to the aquatic environment. Such additional scientific evidence should normally be based on the studies required at Level 1, or studies of equivalent value, and could include:
>
> (i)   a proven potential to degrade rapidly in the aquatic environment;
> (ii)   an absence of chronic toxicity effects at the solubility limit, e.g. where the no-observed effect concentration determined in a prolonged toxicity study with fish or *Daphnia* is greater than the solubility limit.
>
> ## Non-aquatic environment
>
> Substances should be classified as dangerous for the environment, assigned the corresponding symbol (Figure 5.1) with the indication of danger "dangerous for the environment" and at least one of the following risk phrases in accordance with criteria to be defined.
>
> R54:   Toxic to flora
> R55:   Toxic to fauna
> R56:   Toxic to soil organisms
> R57:   Toxic to bees
> R58:   May cause long-term adverse effects in the environment
> R59:   Dangerous for the ozone layer
>
> Substances which on the basis of the available evidence concerning their toxicity, persistence, potential to accumulate and predicted or observed environmental fate and behaviour may present a danger, immediate or long-term and/or delayed, to the structure and/or functioning of natural ecosystems other than those covered above.

[1]The criteria are to be found in the 18th Adaptation to Technical Progress (ATP): Classification as "Dangerous for the Aquatic Environment", published as annexes to 93/21/EEC, *Official Journal* (L110A, 4 May 1993).

Box 5.3   Safety phrases for substances dangerous for the environment specified under 67/548 legislation[1]

S54:   *Obtain the consent of pollution control authorities before discharging to waste water treatment plants*

Applicability and criteria for use:

- Applies to substances which may affect the functioning of sewage treatment plant processes and sludge disposal.
- Recommended for substances which are very toxic, toxic or harmful to aquatic organisms or which may cause long-term adverse effects in the aquatic environment.
- Recommended when such substances are used in industry.

S55:   *Treat using the best available techniques before discharge into drains or the aquatic environment*

Applicability and criteria for use:

- Recommended for substances which are very toxic, toxic or harmful to aquatic organisms or substances which may cause long-term adverse effects for which treatment techniques are available.
- Recommended when such substances are used in industry.

S56:   *Do not discharge into drains or the environment, dispose to an authorised waste collection point*

Applicability and criteria for use:

- Recommended for substances which are very toxic or toxic to aquatic organisms or which may cause long-term adverse effects in the aquatic environment.

S57:   *Use appropriate containment to avoid environmental contamination*

Applicability and criteria for use:

- Recommended for substances which are very toxic or toxic to aquatic organisms and particularly for substances which may cause long-term adverse effects in the aquatic or non-aquatic environment.
- Substances toxic to flora, fauna, soil and other organisms.
- Recommended when such substances are used in industry.

S58:   *To be disposed of as hazardous waste*

Applicability and criteria for use:

- Recommended for substances which are very toxic, toxic or harmful to aquatic organisms or substances which may cause long-term adverse effects in the non-aquatic or aquatic environment.
- Recommended for substances toxic to flora, fauna, bees or other organisms.

S59:   *Refer to manufacturer/supplier for information on recovery/recycling*

Applicability and criteria for use:

- *Obligatory* for substances dangerous for the ozone layer.

(*continued*)

Box 5.3   (*Continued*)

- Recommended for substances which are toxic to flora, fauna, soil organisms, bees or substances which may cause long-term adverse effects in the environment.

S60:   *This material and/or its container must be disposed of as hazardous waste*

     Applicability and criteria for use:

- This phrase should be used in place of S58 in cases where contaminated containers require disposal.
- Recommended for substances which are very toxic, toxic or harmful to aquatic organisms or substances which may cause long-term adverse effects in the non-aquatic or aquatic environment.
- Recommended for substances toxic to flora, fauna, bees or other organisms.

---

[1]Normal working practice under ATP procedures (see footnote 1, Box 5.2) is for all substances classified in respect of the aquatic environment to be given S61, and in addition for R50/53 substances to be given S60. R59 substances receive S59.

What the competent authorities and Commission do with the information

In the case of new substances, the competent authority receiving notification is required to carry out a risk assessment according to the principles laid down in Directive 93/67 which is supplemented by Technical Guidance Notes (now in a consolidated form that refers to both new and existing substances – see section 5.2.4). The first thing to note though is that the procedures will only be carried out on substances classified as dangerous for the environment and so is hazard driven unless, that is, there are other reasonable grounds for concern. These could include:

1. Indication of bioaccumulation potential.
2. The shape of the toxicity/concentration curves in ecotoxicity testing.
3. Indications of other adverse effects on the basis of toxicity studies (e.g. mutagenicity).
4. Data on structurally analogous substances.

In the Directive, risk characterisation (not assessment) is defined as

> the estimation of the incidence and severity of the adverse effects likely to occur in a human population or environmental compartment due to actual or predicted exposure to a substance, and may include "risk estimation", i.e. the quantification of that likelihood.

"Risk assessment", it goes on (Article 3), "shall entail hazard identification and, as appropriate, dose (concentration) – response (effect) assessment, exposure assessment and risk characterisation."

Risk assessments (not characterisations!) are then defined more precisely in annexes. That for the environment makes it clear that as far as the Directive goes, risk characterisation (not assessment) shall consist of comparing predicted exposure concentrations (PECs) with predicted no effect concentrations (PNECs) as discussed in Chapter 2. It is recognised that PNECs very often will have to be assessed on the basis of data from acute tests using assessment factors. In a footnote it says:

> An assessment factor of the order of 1000 is typically applied to an $L(E)C_{50}$ value derived from the results of testing for acute toxicity but that factors may be reduced in the light of more relevant information. A lower assessment factor is typically applied to a NOEC derived from the results of testing for chronic toxicity.

If the PEC/PNEC ratio is equal to or less than 1 then the substance is deemed of no immediate concern and need not be considered again unless and until further information is made available, and the same would be true of substances not classified as dangerous for the environment subject to the provisions mentioned above. However, if the ratio is greater than 1 the competent authority has to decide, on the basis of the size of that ratio and other relevant factors, one of the following conclusions:

- If further information is needed either immediately or at a time when the quantity placed on the market reaches the next tonnage threshold as specified above (p. 61)
- On immediate recommendations for a risk reduction programme

There is an important point about **integration**. The risk characterisation may be carried out to relate to more than one environmental compartment (section 2.4). The competent authority is required to decide which of the above conclusions is applicable to each compartment. Then having completed the risk assessment (not characterisation!) the competent authority shall review the different conclusions and produce an integrated conclusion in relation to the overall environmental effects of the substance.

Notifiers are not themselves formally required to carry out a risk assessment, but clearly would be advised to do so, as they are in the Directive: "the notifier may also provide the authority with a preliminary assessment of the risks" (Article 7(1)).

The dossiers and risk assessments (or summary of them) are forwarded to the Commission which sends copies to Member States. The Commission may then contemplate control on the basis of the risk assessment through, for example, the marketing and use controls described below.

Implementation in Britain

Powers of enforcement with respect to labels, packaging and compliance with the requirements of notification are through the Health and Safety at Work, etc. Act 1974 and using the transposition powers in the European Communities Act 1972. The Health and Safety Executive (HSE) and DoE are designated as formal notifying authorities.

5.2.2 Dangerous preparations

Directive 88/379/EEC on the approximation of the laws, regulations and administrative provisions of the Member States relating to the classification, packaging and labelling of dangerous preparations.

Legal basis: Article 100a.

Aims

Mixtures or solutions of two or more substances (e.g. paints and solvents) are to be classified, labelled and packaged, as with substances (see previous section) on the basis of their hazards.

Details

Any preparations that contain at least one substance classified as dangerous under 67/548 or regarded as dangerous under criteria that are laid down, except medicines, cosmetics, wastes, munitions, explosives, foodstuffs and animal feeding stuffs, which are addressed under other legislation that is concerned with harm to humans, are subject to this Directive. Pesticides are also excluded (see below). The preparations are classified as follows: very toxic, toxic, harmful, etc. Currently, though, the legislation does not include a "dangerous for the environment" category. Neither does it include a general notification requirement. At the time of writing this is under review.

Implementation in Britain is as for dangerous substances (see previous section).

5.2.3 Pesticides

Directive 91/414/EEC concerning the placing of plant protection products on the market.

Legal basis: Article 43.

Aims

This is both a controlling and information-gathering instrument. Plant protection products cannot be placed on the market and used unless the active substances are listed in Annex 1. Article 5 indicates that for inclusion in Annex 1, fate, distribution as well as impact on non-target species, have to be taken into account. Annex II details how application should be made for Annex 1 status.

Details

Two dossiers are submitted for substances proposed for Annex 1 status. One supplies information on the active substances necessary for evaluating foreseeable risks (i.e.

their use, when consistent with good practice, must demonstrably have no adverse effects on human or animal health, on the environment and on non-target species) and a proposed classification following 67/548 (above). The other provides information on one particular formulation of the active ingredient to permit evaluation of effectiveness and foreseeable risk.

The dossier is submitted to competent authorities and then to other Member States and the Commission. The latter refers the dossier to a Standing Committee on Plant Health which advises the Commission on inclusion in Annex 1.

Member States authorise pesticide products (i.e. formulations of active ingredients) according to certain criteria and in particular that active ingredients are in Annex 1.

A key element is the principle of mutual acceptance, under which each Member State should be prepared to accept product approvals granted in other States, unless there are good reasons to differ.

Implementation in the UK

In the UK, several government departments and an independent advisory committee are concerned in the registration of pesticides. The regulatory assessments of products used for plant protection in an agricultural context are largely channelled through the Ministry of Agriculture, Fisheries and Food and non-agricultural pesticides through the HSE.

Related EC legislation

Directive 79/117/EEC **bans** all pesticide products containing substances listed on an Annex – that are there because of recognised risks to man and the environment. The Annex has been amended several times. This negative list contrasts with the positive list established under 91/414. There is also a series of Directives that are not primarily intended to protect the environment but consumers by setting limits on the amounts of pesticides in food (fruit and vegetables, cereals, foodstuff of animal origin) and water. There is a proposal, generally known as the "Biocides Directive", for registration of all non-agricultural pesticides that would operate in a similar way to 91/414/EEC.

5.2.4   Existing substances

Regulation (EEC) No. 793/93 on the evaluation and control of the risks of existing substances.

Legal basis: Article 100a.

Aims

This Regulation is designed to expose information, and gaps in it, on existing chemicals, particularly those identified as being especially worrying through their level of use or intrinsic properties. It will be noted that this legislation is embodied in a Regulation that has direct effect across the EU rather than, as with new chemicals (above), through a Directive that requires implementation through individual legislation. The reasons given for this are that whereas new substances will be manufactured in particular places in the EU so that Member States are more appropriately involved in interacting with manufacturers, existing substances may be *used* throughout the EU. So: "a Regulation is the appropriate legal instrument as it imposes on manufacturers and importers precise requirements to be implemented at the same time and in the same manner throughout the Community".

Details

Which substances?   The Regulation is concerned with substances listed on EINECS. The first priority recognised by the Regulation is the need to address those produced at high volume. Hence any manufacturer or importer who has produced (or imported) > 1000 tonnes p.a. must submit (within 24 months of adoption, i.e. by now) the following:

- Name of substance and EINECS number
- Quantity involved
- Classification as in 67/548 (above)
- Information on the reasonably foreseeable uses of the substances
- Data on the physico-chemical properties
- Data on fate
- Data on ecotoxicity
- Data on acute and subacute toxicity
- Data on carcinogenicity, mutagenicity and toxicity

For manufacturers (importers) involved with production (or importation) levels < 1000 tonnes p.a. but > 10 tonnes p.a., the following must be supplied within a period of four years of the coming into force of the Regulation:

- Name and EINECS number
- Quantity produced or imported
- Classification as in 67/548

Note that in the context of the Regulation "production" means that the substance becomes physically isolated in a process; it is not just an intermediary, but it does not just include **useful** products.

The Commission in consultation with Member States may require more information.

In all cases manufacturers and importers must make all reasonable efforts to obtain existing data but are not bound to carry out further tests at this stage.

The data are submitted on computer diskette as specified in Annex III, and this is referred to as a harmonised electronic data set (HEDSET). This is believed to be the first time that a requirement for "electronic data collection" has been made in any EC legislation. For the high production volume chemicals the Commission received more than 10 000 diskettes on a total of *c*. 1400 substances.

What the Commission does with the information: Prioritisation.   On the basis of the information submitted, the Commission is required regularly to draw up priority lists of substances needing immediate attention. Over the first couple of years these were based on suggestions made by Member States. However, routinely prioritisation will be done on the basis of an automated, score-based computer assessment of data supplied by manufacturers and importers and taking into account effects, exposure, data gaps, work already carried out in other international forums (see later), other Community legislation and/or programmes (see Box 2.3). The method is called the informal working group on priority setting (IPS) method. Before selecting substances for the finalised priority tests, expert judgement enters the process, by giving the Member States, industry and selected NGOs an opportunity to comment on the IPS ranking. The revised ranking, known as the formal group on priority setting (FPS) ranking, is then used to select the priority substances.

What happens to substances on priority lists?   The Commission is required to publish (in the *OJ*) priority lists after adoption. For substances on the lists, responsibility for carrying out a risk characterisation is shared out between Member States through designated rapporteurs. Manufacturers and importers are required, within six months of the list being published, to give the rapporteur all the relevant information for risk assessment of the substance concerned. This obligation formally falls on **all** manufacturers and importers who have submitted data as part of the prioritisation exercise. The data to be made available have to at least include the base set as required under 67/548/EEC. Exemptions are possible but must be acceptable to other Member States, and are also forwarded to the OECD so that non-EU member countries have a chance to comment. The rapporteur may at this stage require more testing in accordance with the general principles already outlined. The rapporteur, as for the competent authority in 67/548, is required to evaluate risk of the substance to man and the environment according to procedures defined in a daughter Regulation ((EC) No. 1488/94) and the associated, consolidated Technical Guidance Notes. The principles are similar to those already discussed for new substances except they recognise that for existing substances not all available ecotoxicity data will be from standard tests carried out according to GLP, and there may be data on actual exposure concentrations and effects from field monitoring programmes. Draft reports are sent to the Commission for comments, not only by other Member States but also to the OECD for consideration by non-EU member countries, and if the substance is considered to be a problem in the workplace by an advisory committee of DG V (concerned with employment, industrial relations and social affairs).

When one or more of the PEC/PNEC ratios (for different geographical scales and environmental compartments; section 2.4) exceed 1 the rapporteur is also required to suggest a strategy for risk management (see Chapter 3) that will be guided in terms of where it is directed and how it is effected by analyses of the PEC/PNEC ratios. Thus if only local ratios were signalling problems, management could be directed at particular sites or industries. But if regional or global ratios were signalling problems restrictions on marketing and use may be more appropriate. In the latter case the rapporteur is also required to submit an analysis of the advantages and drawbacks of the substances and of the availability of replacement substances. This is essentially a cost–benefit analysis and we shall return to it in Chapter 8. In assessing both candidates for priority lists, and recommendations on management strategies, the Commission is assisted by a committee composed of representatives of the Member States. All information is treated as public, unless manufacturers/importers can establish the need for commercial confidentiality.

Implementation in the UK

This being a Regulation does not require implementation through the law of individual Member States. However, some national legislation is necessary to provide powers for enforcement, in terms of requiring manufacturers/importers to provide necessary information and carry out supplementary tests, and to provide penalties for failing to comply with the requirements. This could be done through EPA 90, Part VIII, section 142. However, since this legislation involves all substances on EINECS, including pesticides, it goes beyond the terms of EPA 90 and so powers are given through a statutory instrument under the European Communities Act 1972. The enforcement authorities are the Environment Agency and HSE.

5.3   CONTROLLING LEGISLATION

Here, as has been done before, a distinction is made between pollution control as applied to point sources and the more distributed problems of controlling marketing and use of substances.

5.3.1   Pollution controls for (largely) point sources in aquatic systems

---

Directive 76/464/EEC on pollution caused by certain dangerous substances discharged into the aquatic environment.

Legal basis: Articles 100 and 235.

Aims

This Directive has two major aims: first, and foremost, to provide a framework for the elimination or reduction of pollution of inland, coastal and territorial waters by particularly dangerous substances; but second, as the legal basis of the Directive implies, to avoid the creation of unequal conditions of competition. It is important to note that pollution is defined as "the discharge by man, directly or indirectly, of substances or energy into the aquatic environment, the results of which are such as to cause hazards to human health, harm to living resources and to aquatic ecosystems, damage to amenities or interference with other legitimate uses of water". Hence, the legislation does not advocate zero emissions, but aims to achieve emissions likely to have zero effects.

Details

The Directive establishes two lists of substances for action. List I – sometimes referred to as the Black List – a group of dangerous substances, pollution from which has to be *eliminated*; List II – sometimes referred to as the Grey List – a group of substances for which it is necessary to reduce emission in order to avoid water pollution. The framework Directive does not specify controls associated with particular substances. Instead an Annex contains a list of candidates for List I and II substances (Table 5.1). Daughter directives are intended to define the controls for specific substances on List I. Candidate substances on List I are to be treated, in terms of controls, as if they were List II substances (see below) until a specific daughter directive is produced on them.

The Annex to the Directive contains only a list of families and groups of substances for consideration. This was refined in 1982 by a Communication from Commission to Council listing 129 as priority candidate substances. The Communication emphasised that the list was not final and the Commission subsequently added another three substances.

Controls applied to List I and II substances

Pollution from List I substances has to be eliminated. But Member States can choose between two alternative regimes. One by **limit values** for which **emission standards** set at national levels must not exceed (but they can be more stringent). The limit values are fixed uniformly throughout the EU in the daughter directives. The alternative involves **emission standards** set by reference to **quality objectives,** also laid down in the daughter directives. The use of the latter is conditional on Member States proving to the Commission that the quality objectives (UK terminology refers to these as quality standards) are being met in accordance with a monitoring procedure set up by the Council.

For List II substances Member States have to establish pollution reduction programmes with deadlines for implementation. All discharges liable to contain List

Table 5.1   The European Commission priority candidate list; known as the Black List

2 2-Amino-4-chlorophenol	65 1,2-Dichloropropane
3 Anthracene	66 1,3-Dichloropropan-2-ol
5 Azinphos-ethyl	67 1,3-Dichloropropene
6 Azinphos-methyl **R**	68 2,3-Dichloropropene
8 Benzidine	69 Dichlorprop
9 Benzyl chloride	70 Dichlorvos **R**
10 Benzylidene chloride	72 Diethylamine
11 Biphenyl	73 Dimethoate
14 Chloral hydrate	74 Dimethylamine
16 Chloroacetic acid	75 Disulfoton
17 2-Chloroaniline	76 Endosulfan **R**
18 3-Chloroaniline	78 Epichlorohydrin
19 4-Chloroaniline	79 Ethylbenzene
21 1-Chloro-2, 4-dinitrobenzene	80 Fenitrothion **R**
22 2-Chloroethanol	81 Fenthion
24 4-Chloro-3-methylphenol	86 Hexachloroethane
25 1-Chloronaphthalene	87 Isopropylbenzene
26 Chloronaphthalenes (technical mixture)	88 Linuron
27 4-Chloro-2-nitroaniline	89 Malathion **R**
28 1-Chloro-2-nitrobenzene	90 2-Methyl-4-chlorophenoxyacetic acid
29 1-Chloro-3-nitrobenzene	91 2-Methyl-4-chlorophenoxypropanoic acid
30 1-Chloro-4-nitrobenzene	93 Methamidophos
31 4-Chloro-2-nitrotoluene	94 Mevinphos
32 Chloronitrotoluenes (other than 4-chloro-2-nitrotoluene)	95 Monolinuron
33 2-Chlorophenol	96 Naphthalene
34 3-Chlorophenol	97 Omethoate
35 4-Chlorophenol	98 Oxydemeton-methyl
36 Chloroprene	99 PAH (with special reference to 3,4-benzopyrene and 3,4-benzofluoranthene)
37 3-Chloropropene	100 Parathion (including parathion-methyl)
38 2-Chlorotoluene	101 PCBs (including PCTs) **R**
39 3-Chlorotoluene	103 Phoxim
40 4-Chlorotoluene	104 Propanil
41 2-Chloro-p-toluidine	105 Pyrazon
42 Chlorotoluidines (other than 2-Chloro-p-toluidine)	106 Simazine **R**
43 Coumaphos	107 2,4,5-T (including salts and esters)
44 Cyanuric chloride	108 Tetrabutyltin
45 2,4-D (including salts and esters)	109 1,2,4,5-Tetrachlorobenzene
47 Demeton (including demeton-0; -S; -S-methyl; -S-methyl sulphone)	110 1,1,2,2-Tetrachloroethane
48 1,2-Dibromoethane	112 Toluene
49 Dibutyltin dichloride	113 Triazophos
50 Dibutyltin oxide	114 Tributyl phosphate
51 Dibutyltin salts (other than dibutyltin chloride and dibutyltin oxide)	115 Tributyltin oxide **R**
52 Dichloroanilines	116 Trichlorfon
53 1,2-Dichlorobenzene	119 1,1,1-Trichloroethane
54 1,3-Dichlorobenzene	120 1,1,2-Trichloroethane
55 1,4-Dichlorobenzene	122 Trichlorophenols
56 Dichlorobenzidines	123 1,1,2-Trichlorotrifluoroethane
57 Dichlorodiisopropyl ether	124 Trifluralin **R**
58 1,1-Dichloroethane	125 Triphenyltin acetate
60 1,1-Dichloroethylene	126 Triphenyltin chloride
61 1,2-Dichloroethylene	127 Triphenyltin hydroxide
62 Dichloromethane	128 Vinyl chloride
63 Dichloronitrobenzenes	129 Xylenes (technical mixture of isomers)
64 2,4-Dichlorophenol	131 Atrazine **R**
	132 Bentazone

R = on UK Red List, which also contains other substances including pesticides.

II substances require prior authorisation by the competent authority in which emission standards are to be laid down. These standards are to be based on the achievement of quality objectives.

There was considerable controversy associated with deciding between the use of limit values and quality objectives (standards) for List I substances, with the UK holding out against all the other Member States for the use of the latter in preference to the former. So a compromise, the parallel approach, emerged. The differences between the two approaches were discussed in Chapter 3 and are that:

- Limit values are fixed (often for particular industrial sectors) and are applied "globally" with no account taken of local conditions.
- Quality objectives (standards) are more flexible, allowing total inputs from both point and diffuse sources to be taken into account when emission standards are set.

Limit values are administratively more convenient and are less likely to lead to distortions in competition than quality objectives. The quality objectives are ecologically more defensible. There is obvious inconsistency in the way List I and II substances are treated.

Finally, it is worth repeating that candidate List I substances are treated as List II substances, until they are given confirmed List I status.

Steps in establishing controls on List I substances

There are four technical stages in setting controls for List I substances:

1. Selection of candidate list.
2. Prioritisation of candidates for detailed analysis.
3. Confirmation of a particular candidate for List 1 status.
4. Establishment of controls.

In **stage 1** the Commission initially selected 500 substances produced in quantities of more than 100 tonnes p.a. across the EU, and then on the basis of advice from consultants on hazards to the aquatic environment selected the initial list of 129 (Table 5.1). But the Directive makes clear that the Commission acting on its own initiative or at the request of a Member State, can propose revisions and supplements to the Council.

In **stage 2** the Commission selects substances for further consideration on the basis of criteria that are not altogether clear but which take account of the likelihood that substances will be present in EU waters at levels that cause particularly important environmental problems. The UK has been somewhat critical of this and developed its own Red List of 23 (Table 5.1) of the most dangerous chemicals from the list of 129 by explicit prioritisation procedures, amongst other things to focus Commission thinking in this area.

In **stage 3** consultants are commissioned to collect appropriate data on the prioritised candidates and these are then considered by the Commission's Scientific

Advisory Committee on Toxicity and Ecotoxicity of Chemicals. The analysis is hazard based, but in line with the requirements of the Directive takes into account not only toxicity but also persistence and bioaccumulation potential. The rules the Committee operates with respect to ecotoxicity are that if the $L(E)C_{50}$ derived from tests with aquatic organisms $\leqslant 10\,mg/l$ then the substance is confirmed for List I. If substances have $L(E)C_{50} > 10\,mg/l$ they are not normally considered for inclusion provided that they can be degraded to a harmless substance within a few (to up to three) days, the bioaccumulation potential is not a cause for concern and there is no evidence for carcinogenic or mutagenic effects. If any of these supplementary conditions does not apply then the substance is reconsidered for inclusion in List 1.

In **stage 4**, the Commission again seeks advice from the Advisory Committee, but exclusively with regard to the quality objectives. Here, the procedures adopted by the Committee are similar to those used in the classification procedures already described (p. 59). It is hazard based and emphasises ecotoxicity, with information on persistence and bioaccumulation acting as modifying factors. The aim of the quality objective is to specify the maximum amount of chemical which may be present without causing harm to the aquatic ecosystem; in other words the ecosystem NOEC (see above, p.19). However, rarely are whole ecosystem data available, and most often only acute data are available. So the Committee uses extrapolation factors to obtain an estimate of these ecosystem level responses, higher factors being applied with less certain data:

- Dividing by 1000, where the data available are few and for acute tests or the range of organisms on which results are available is narrow.
- Dividing by 100, where there is a more extensive database of $L(E)C_{50}s$ covering a wider range of species – or where one or a few chronic results are available.
- Dividing by 10, where there are a larger number of representative chronic tests.

Of course, judgement must be applied in this procedure and this is the *raison d'être* of the Advisory Committee.

The Committee transmits its views in the form of opinions. On the basis of this advice, the Commission drafts daughter directives that specify the maximum concentration of a substance permissible in a discharge and, where appropriate, the maximum quantity of such a substance (unit weight) per unit of raw material or per product. These limit values can be established according to sector and type of product. As well as taking into account toxicity, persistence and bioaccumulation they should also take account of best technical means (see below). The draft also lays down quality objectives. These are submitted to Council and a decision is taken on the basis of unanimous voting (Article 12).

Daughter directives

The parent Directive was adopted in 1976, but by 1992 only two daughter directives had been agreed. This was largely due to extensive debate between Member States on the relative merits of the different approaches, the implications for both competition and the environment and, of course, the requirement for unanimity. Another point of discussion was whether new plant should be required to use best available technology for limiting emissions, even where the quality objective approach was being followed. In an attempt to short-circuit these discussions a Daughter Directive 86/280/EEC was adopted that contains a standard set of clauses: authorisations are generally to conform with limit values or quality objectives; limit values in terms of concentration should not in principle be exceeded, but those in terms of quantity must be observed; for quality objectives it is for Member States to determine areas affected by discharges; over and above these controls that apply to point sources, Member States are required to draw up programmes to avoid or eliminate pollution from multiple and diffuse sources; general principles for monitoring and analysis are defined.

As far as new plant and best technical means are concerned, Article 3, para. 4 of 86/280/EEC states that "whatever method it adopts, the Member States . . . concerned shall, where for technical reasons the measures envisaged do not correspond to best technical methods available, provide the Commission, before any authorisation, with evidence in support of these reasons".

The Daughter Directive 86/280/EEC did speed up the process; before adoption only three substances had been controlled, after adoption 14 substances were made the subject of daughter directives in four years. Prior to the adoption of the facilitating Directive the daughter directives were agreed separately. Thereafter they were developed in the framework of 86/280/EEC. The directives are listed in Table 5.2. Another measure of progress, though, comes from the realisation that though the Scientific

Table 5.2 Daughter Directives to 76/464

	Substances	Directive
1	Mercury	82/176/EEC and 84/156/EEC
2	Cadmium	83/513/EEC
3	Hexachlorocyclohexane	84/491/EEC
4	DDT	
5	Carbon tetrachloride	86/280/EEC
6	Pentachlorophenol	
7	Aldrin	
8	Dieldrin	
9	Endrin	
10	Isodrin	88/347/EEC
11	Hexachlorobenzene	
12	Hexachlorobutadiene	
13	Chloroform	
14	1,2-Dichloroethane	
15	Trichloroethane	90/415/EEC
16	Perchloroethane	
17	Trichlorobenzene	

Advisory Committee has given advice on quality objectives for approximately 50% of presently existing List I substances, only 17 have been presented to and accepted by the Council, while 16 are currently tabled for consideration.

Implementation in the UK

The main UK instruments of Directive 76/464 and its daughters are now the Water Resources Act 1991 and Environmental Protection Act 1990. These establish consents and authorisations that ensure conformity with the quality objectives (standards) and the limit value requirements. Regulatory authorities were the NRA and HMIP, but now their roles have been taken over by the Environment Agency.

Environmental quality standards and emission standards for both List I and II substances are currently being established by the DoE. These are progressively given statutory force through regulations under the Water Resources Act 1991. These are subject to public comment. All representations and objections have to be considered and proposals may be modified in the light of them.

The Water Industry Act 1991 also gives powers to regulate discharges to sewers of List I substances. The responsibility for consents is initially with the sewerage under-taker (except for Red List substances) which is a unique situation in Britain: it is the only instance of a private body exercising regulatory functions in the context of envir-onmental protection. But the NRA and now the Environment Agency are involved, and the presence of Black and Grey List substances in the final outfall of a sewage works must be authorised by the Environment Agency in a way that ensures compli-ance with environmental quality requirements.

The relationship between environmental standards and the application of best available technology is also being explored by the Environment Agency. The quality standard defines a level above which contamination is not tolerable. Below this there is a range of concentrations that are associated with tolerable risks. It is in this range that controls can be implemented by the exercise of best available techniques. Below this level risks are so low as to be judged as minimal. Here best available techniques can be applied, but the cost of implementation should be taken more seriously (i.e. BATNEEC). This is discussed further in Chapter 8.

5.3.2   Pollutant controls for (largely) point source emissions to atmosphere

Directive 84/360/EEC on combating of air pollution from industrial plants.

Legal basis: Articles 100 and 235.

Aims

Initiated largely as a result of concerns about the effects of atmospheric pollution, especially acid deposition on forests, this Directive establishes a framework for more general controls on emissions from industrial plants.

Details

The purpose of this Directive is to prevent or reduce air pollution from industrial plants within the EU, especially those listed in an Annex and in the following major categories: energy industry; production and processing of metals; manufacture of non-metallic mineral products; chemical industry; waste disposal; manufacture of pulp by chemical methods. Plants belonging to these categories require prior authorisation by competent authorities in the Member States either at design stage or in the case of substantial alterations.

Authorisation can be given only when all appropriate prevention measures have been taken, including the application of best available technology provided the application of such measures does not entail excessive costs (this, in fact, is the source of BATNEEC – see Chapter 1). The competent authority has to be satisfied that the plant will not cause significant air pollution, especially from substances listed in a second Annex: sulphur dioxide and its compounds; oxides of nitrogen and other nitrogen compounds; carbon monoxide; organic compounds, in particular hydrocarbons (except methane); heavy metals and their compounds; dust (asbestos (suspended particulate fibres), glass and mineral fibres); chlorine and its compounds; fluorine and its compounds. Finally, emission limit values and air quality limit values have to be specified. These may be as fixed by the Council, but Member States have some flexibility, i.e. defining particularly polluted areas for which more stringent emission limit values may be applied and defining specially protected areas for which more stringent air quality limit values and emission limit values may be imposed. The Council can also lay down measurement and assessment techniques and methods.

So, as with the dangerous substances Directive relating to the aquatic environment, this approximates to a framework directive in that it presumes subsequent directives setting emission limit values. These in fact have been set in subsequent directives as follows:

- *Asbestos* (87/217/EEC) – this places a general duty on Member States to ensure that emissions not only to air but also land and water (i.e. integrated pollution control in embryo) are, as far as reasonably practicable, reduced at source and prevented. BATNEEC is referred to.
- *Large combustion plants* (88/609/EEC) – this lays down emission limit values for SO_2 and NO_x from fossil-fuelled power stations and other large combustion plants such as oil refineries. Different requirements are set for new and existing plant; and for the latter these are specified as total national emission limits with phased reduction and with different limits for different Member States.

- *Municipal waste incineration* (89/369/EEC and 89/429/EEC) – this lays down emission limit values and operating and monitoring requirements. The emission limit values cover: dust; heavy metals; acid gases (HCl, HF and SO_2). Member States may also fix emission limits for other pollutants such as dioxins and furans. Maximum concentrations of CO and organic compounds in combustion gases are also specified.

It should also be noted that two directives limit the lead content in petrol (85/210/EEC and 87/416/EEC), and a series of directives set limits on CO, hydrocarbon, NO_x, VOCs (see Glossary) and particulates from passenger cars (70/220; 72/306; 74/290; 77/102; 78/665; 83/351; 88/76; 88/436; 89/458; 89/491; 91/441; 94/12), commercial vehicles (88/77; 91/542) and diesel engines for tractors (77/537).

Air quality standards

These directives, that predate 84/360, establish air quality standards for: smoke and sulphur dioxide; nitrogen dioxide; lead. These will be dealt with briefly and in turn below:

- *Smoke and sulphur dioxide* (80/779/EEC) – this sets limit values for the ground-level concentration of SO_2 and suspended particulates to protect human health and the environment. These limit values for the two pollutants are considered together; if one is low, more is allowed of the other. This is dubious practice since the synergy of effect that the practice implies is not obvious, especially from an environmental perspective. Measuring stations have to be established.
- *Nitrogen dioxide* (85/203/EEC) – this sets an air quality standard designed to protect human health and to contribute towards protection of the environment. Measuring stations have to be established.
- *Lead* (82/884/EEC) – this sets a limit value for lead in the air largely to protect human health.

Implementation within the UK

Part I of the Environmental Protection Act 1990 was written, to a significant extent, to address the requirements of 84/360: it lists prescribed processes that are the same as those in the Directive, with some additions; there is a requirement for authorisation based on BATNEEC. The Environment Act 1995 puts a duty on the Secretary of State to develop a National Air Quality Strategy setting air quality standards (AQSs) – e.g. for NO_x and SO_2, etc. – and puts a duty on local authorities to develop air quality management areas (AQMAs) in places where objectives are not being met. Prior to this, controls have been effected through the European Communities Act 1972, the Health and Safety at Work Act 1974 and the Environmental Protection Act 1990.

5.3.3   Controls on marketing and use; a more distributed problem

> **Directive** 76/769/EEC on restrictions on the marketing and use of certain dangerous substances and preparations (initially PCBs, PCTs and monomer vinyl chloride).
>
> **Legal basis**: This parent Directive and the first eight amending directives refer to Article 100, subsequent amendments refer to 100a.

Aims

This Directive aims to control the marketing and use of dangerous substances and preparations. Restrictions are specified in an Annex of the framework Directive. Subsequent daughter directives have extended the Annex. As the legal basis implies, the initial impetus was to prevent controls on marketing and use influencing free trade within the common market. However, "the rules concerning the placing on the market of dangerous substances and preparation must aim at protecting the public, and in particular persons using such substances and preparations...they should contribute to the protection of the environment".

Details

The details of the various amendments that add substances to the Annex of 76/769 are summarised in Table 5.3. They involve restrictions on the marketing and/or use of a range of substances, often with derogations. The legislation is scattered through the *OJ*s and a proposal for a consolidated version is currently under consideration.

As already noted, the initial motivation for the Directive was to prevent distortions in the free market. So it is not surprising that much of the legislation has been involved with catching up with measures already taken by individual Member States, often based upon suspicion of pollution or hazard (p. 5). Increasingly, though, it is anticipated that substances will fall into the compass of the marketing and use Directive on the basis of risk assessments carried out in the context of existing and new substances legislation. The effect of basing legislation on Article 100a means that it can be adopted by qualified majority voting but in principle cannot be applied more rigorously by Member States. In addition, though, Article 100a allows Member States to adopt higher national standards relating to the protection of the environment if these are not means of arbitrary discrimination (e.g. German and Danish response to the Ninth Amendment on PCPs; p. 50).

Implementation in the UK

The marketing and use of chemicals involves potential interfaces with people at home, at work and in the environment as well as on the environment itself. It should not be

Table 5.3 EU and UK legislation under Directive 76/769

EU legislation	What it does	UK Regulations (SIs)
76/769 parent, framework	PCBs and PCTs only to be used in closed-system electrical equipment, large condensers and other specified applications. Monomer vinyl chloride banned as aerosol propellant	Control of Pollution (Supply and Use of Injurious Substances) (CP (S and UIS)) [1980]
79/663 1st amendment	Trichloroethylene, tetrachloroethylene, carbon tetrachloride banned from use in ornamental objects. Tris banned from use as fire retardant in textiles that make contact with skin	Dangerous Substances and Preparations (DS&P) (Safety) (Consolidation) [1994]
82/806 2nd amendment	Bans use of benzene in toys in excess of 5 mg/kg toy	DS&P (Safety) (Consolidation) [1994]
82/828 3rd amendment	Relaxed 76/769 by extending time limit for use of PCTs in reusable thermoplastic tooling (to 31 December 1984)	CP (S and UIS) [1986]
83/264 4th amendment	Bans two substances used as fire retardants and three substances used in sneezing powders and in novelties, jokes and hoaxes	DS & P (Safety) (Consolidation) [1994] and Novelties (Safety) [1985]
83/478 5th amendment	Restricts marketing and use of various types of asbestos	The Asbestos (Prohibition) [1992] and Control of Asbestos at Work [1987] and Asbestos Products (Safety) [1985]
85/467 6th amendment	Bans all new uses PCBs and PCTs and terminates exemptions permitted by 76/769. Also banned is second-hand marketing of equipment containing PCBs and PCTs	CP (S and UIS) [1986]
85/610 7th amendment	Further restricts marketing and use of asbestos	The Asbestos (Prohibition) [1992] and Asbestos Products (Safety) [1987]

(*continued*)

Table 5.3   *(Continued)*

EU legislation	What it does	UK Regulations (SIs)
89/677 8th amendment	Bans use of: lead carbonates and sulphates in paints; mercury compounds used in various ways; arsenic compounds used in various ways; organostannic compounds used in various ways; DBBs; all with some exemptions	Environmental Protection (EP) (Controls on Injuries Substances) [1992] and Control of Substances Hazardous to Health [1991] and Control of Pesticides [1986] and Control of Pollution (Anti-Fouling Paints and Treatments) [1987]
91/173 9th amendment	Bans use of PCP with exemptions	EP [1993] and Control of Pesticides [1986]
91/338 10th amendment	Bans use of cadmium to colour, stabilise or plate and marketing of such products with exemptions, e.g. for safety	EP [1993]
91/339 and 91/659 11th amendment	Bans marketing and use of PCB substitutes	EP [1992] and the Asbestos (Prohibition) Regulations [1992]
94/27 12th amendment	Bans nickel in costume jewellery	?
94/48 13th amendment	Bans certain flammable substances in aerosols intended for entertainment	?
94/60 14th amendment	Pot-pourri dealing with: 1. Substances that are carcinogenic, mutagenic or toxic to reproduction 2. Creosote 3. Certain chlorinated solvents	?

surprising, therefore, that a wide range of UK legislation has been used to provide powers for implementation of the requirements of the directives under regulations. The regulations are summarised in Table 5.3 and involve, for example, all the legislation referred to earlier in section 5.1. Of particular note are that environmental issues were handled by COPA up to the Seventh Amendment. Thereafter, the Environmental Protection Act 1990 has been used − Part VIII under section 140 in that Act provides the Secretary of State with powers to control the importation, storage, marketing and use of hazardous substances and articles and was specifically written to dovetail with 76/769.

Controls on specific substances

Controls on marketing and use of specific substances involve those on pesticides described earlier, and those concerned with banning the marketing of detergents likely to cause foaming in receiving waters. On the latter, Directive 73/404/EEC prohibits the marketing of specified detergents that do not achieve prescribed limits of biodegradability as judged by test procedures specified in Directive 73/405/EEC for anionic detergents and Directive 82/242 for non-ionic detergents. The latter also amends 73/404 by establishing a Committee for adapting the Directive to technical progress. Implementation in UK law has largely been through the European Communities Act 1972.

5.3.4   Waste

Some mention should also be made of controls on waste, since whether or not this is hazardous will depend, to a large extent, on the chemicals it contains. There is a plethora of legislation on waste too extensive and specialist to deal with here. Of special note though is the waste framework Directive 75/442 (as amended 91/156) that:

- Encourages the prevention or reduction of waste production and its harmfulness, particularly through the development of clean technologies.
- Encourages the recovery of waste, including recycling, reuse or reclamation and the use of waste as an energy source.
- Ensures that waste is recovered or disposed of without endangering human health and without using processes and methods that could harm the environment.
- Establishes an integrated and adequate network of disposal installations taking account of the best available technology not involving excessive costs.

Many of these ideas have been taken further in Britain in EPA 90 Part II. The main thrusts of the legislation are that the producer, and indeed anyone who deals with the waste at any stage, is given a duty of care on the waste, and a licensing system is defined to ensure strict management of the material from source to ultimate end. Especially hazardous wastes are subject to the more stringent provisions of a separate Directive 91/689 that are also addressed in EPA 90 Part II, and incineration of this material is addressed in Directive 94/67/EC. Special provisions are made for the disposal of PCBs (Directive 76/403/EEC) and waste oils (Directive 75/439/EEC as amended by Directive 87/101/EEC).

5.3.5   Major accidents

Also noteworthy is Directive 82/501 (as amended by 87/216 and 88/610) on the major accident hazards of certain industrial activities. This addresses exceptional risks such as massive emissions of dangerous substances where an activity gets out of control in a fire or explosion. This is commonly referred to as the Seveso Directive after the accident in Italy in 1976 that prompted the legislation. It requires manufacturers using certain dangerous substances above certain threshold quantities to limit their consequences for humans and the environment and to report major accidents. More specifically: manufacturers must produce a safety report and on-site emergency plans; a competent authority must produce an off-site emergency plan; the public must be informed of safety measures and of the correct behaviour in the event of an accident.

British regulations implementing the Directive were under both the Health and Safety at Work Act, etc. 1974 and the European Communities Act 1972 (to cater for the environmental aspects) as the Control of Industrial Major Accident Hazards (CIMAH) Regulations. Also of relevance are the Notification of Installations Handling of Hazardous Substances (NIHHS) Regulations; while having a separate purpose these did help locate sites falling within the provision of CIMAH. The competent authority for CIMAH is the HSE, which though not originally intended to deal with the environment is required to do so under the terms of this legislation. The competent authorities for NIHHS are hazardous substances authorities (including district councils, metropolitan authorities, London boroughs, county councils) and HSE has a consultation role.

The Seveso Directive aims at providing the public with safety measures and action needed in the event of emergencies and is currently under revision.

5.4   HOW IT FITS TOGETHER

It will now be clear that there is much complexity and diversity in the chemical controls applied in Europe. There is, nevertheless, some logic to what is done in that the existing and new substances legislation is supposed to generate information that can be a source of inspiration for both point source and marketing and use controls. Risk assessment plays a central role in defining the environmental quality standards and in suggesting the types and intensity of controls applied to the sale, use and disposal of substances. This is why the current regulatory era was described as risk based in Chapter 1. Moreover, as far as existing substances go there is even a requirement to consider lost social benefit from the controls being contemplated and we shall return to this balancing-based approach in Chapter 8.

Yet not all routes to controls necessarily pass through the specific information-gathering instruments associated with new and existing substances legislation and the counterparts for pesticides, biocides and detergents. Another important route has been Directive 83/189/EEC as amended by Directive 94/10/EC, laying down a procedure for the provision of information in the field of technical standards. This requires that Member States tell the Commission about any of their intended legislation on controls that are not covered by existing or intended EC legislation. The

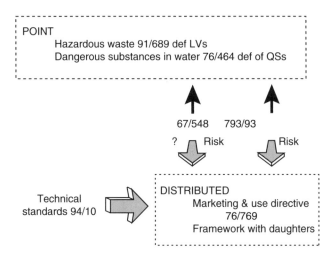

Figure 5.2   How the EC legislation on controlling chemicals fits together

Directive then requires a "standstill" period to enable the Commission and Member States to object or to form catch-up legislation for the EU as a whole. The basis of the legislation has been Article 100 (100a) of the Treaty of Rome, so, not surprisingly, "catch up" has been a preferred option and this has been a major basis for marketing and use controls prior to the adoption of the Existing Substances Regulation. In the past there has been no requirement for Member States to justify the proposals on the basis of any specific requirements and so this has been a source of controls not subject to the rigour of risk assessment. But the amending Directive requires that when national legislation seeks to limit the marketing and use of a chemical substance, preparation or product on the grounds of public health, or protection of consumers or the environment, Member States shall supply to the Commission an analysis of risks carried out in accordance with the requirements of existing and new substances legislation (see above).

Finally, note should be made of the difference between powers provided by the technical standards Directive and paragraph 4 of Article 100a. The former is largely concerned with new controls; the latter with the extension of existing ones (p. 81). Figure 5.2 provides a rough outline of the interrelationships between some of the major pieces of EC legislation described above.

5.5   TOWARDS MORE INTEGRATION

It will be clear from the foregoing that most of the controls on chemicals are targeted at particular substances and separate environmental sectors. Under the Fifth Action

Programme (Chapter 4) the concept of integrated pollution control is given priority in terms of action which must be taken for more rational and hence sustainable environmental protection. The environment cannot be rigidly partitioned, for measures taken in one sector may have implications for others. The aim is to provide controls that achieve the optimum protection for all sectors.

Roots of this kind of thinking are present, as already noted, in the asbestos Directive (87/217; p. 79) that influenced development of the UK Environmental Protection Act, especially Part I. But otherwise it has not been widely involved in EC legislation. However, now there are Commission proposals in the pipeline for a more integrated approach. An integrating framework directive is being developed for the water environment that recognises the interrelationship between the ecological quality of surface waters and groundwater, the quality of effluents and the quality of drinking water. Another piece of legislation (recently adopted; September 1996) is integrated pollution prevention and control (IPPC) for certain industrial processes. This legislation is broadly similar to the UK IPC specified in EPA 90, but is not identical. Its declared objective is to provide for measures and procedures to prevent, or where that is not practicable reduce, industrial emissions within the EU so as to achieve a high level of protection for the environment **as a whole**. These include the establishment of a permit system for installations carrying out activities and processes divided into the same categories as those set out in UK legislation. BATNEEC will be applied and the Commission will set up an information-exchanging arrangement between competent authorities on this. All permit applications and results from monitoring under the permits would be made public – subject to confidentiality requirements.

The trend towards integration is also being developed by two voluntary programmes: one that addresses processes and the other products. These are essentially market-based tools of control.

The process-based instrument is defined in Council Regulation (EEC) No. 1836/93 **allowing voluntary participation by companies in the industrial sector in a Community eco-management and audit scheme** (legal basis: Article 130s). Participation is currently restricted to companies performing industrial activities. The overall objective is to promote continuous environmental performance improvements within industry by committing **sites** to:

- Establish and implement environmental policies, programmes and management systems
- Periodically subject them to audits
- Provide environmental performance information to the public

The operators have to demonstrate compliance with appropriate existing legislation and identify environmental effects of the overall operation and demonstrate improvements. On this basis they receive statements of participation. Hence the system requires an integrated approach to the environmental impact of all the processes and associated outputs emanating from a site.

The products-based instrument is defined in Council Regulation (EEC) No. 880/92 (legal basis: Article ...130s) as a **Community Ecolabel Award Scheme**. The objectives of the scheme are to promote the design, production, sale and use of products less harmful to the environment **throughout their life-cycles** and to inform consumers better about product impact on the environment. This ecolabel (Figure 5.3) is attached to products that have less of a negative impact on the environment, while remaining as effective as comparable products. Product groups are identified and judgements are made with respect to energy and materials used, and environmental impact throughout their life-cycle, i.e. from raw materials, through production, to use and ultimate disposal. Life-cycle assessment (LCA) thus takes centre stage. Hence, in principle, the system requires an integration of all environmental impacts emerging from a class of products. In the UK the system is administered by the UK Ecolabelling Board.

Clearly, both these systems should converge and indeed should take account of each other in the process. In practice, though, audits are more likely to be applied to existing operations, whereas though LCAs will be applied to existing products they will increasingly be used to design new products and hence the processes and plant that go with them.

The main engines of action in both these schemes are customers/public pressure on the basis of information fed into the market-place. Indeed more generally, in order to make use of the same forces, information on environmental issues kept by public bodies is made more publicly available through the **Directive on the freedom of access to information on the environment** (90/313). More recently there has

Figure 5.3   EC ecolabel

been discussion on the possibility of the use of a chemical release inventory in Europe. These have been used to great effect in the USA (p. 101).

5.6 RÉSUMÉ

- The legislation concerning hazardous substances now largely has its origins in the EU – it is complex and wide-ranging, so it is not easy to get an overview.
- Broadly speaking it divides into that concerned with gathering information for prioritisation, classification and risk assessment and that aimed at risk reduction through controls on emissions, marketing and use.
- Information-gathering procedures are now largely based on a rudimentary form of risk assessment and are supposed to provide a basis for controls.
- Risk reduction is achieved mainly by end-of-pipe restrictions, and limitations on importation, storage, marketing and use, and disposal.
- However, there are moves towards more integrated approaches with attention focused both on industrial processes and products. There is also increasing interest being shown in the involvement of market instruments as methods of achieving environmental protection.

5.7 FURTHER READING

Haigh, N. (1995). *Manual of Environmental Policy: the EC and Britain.* Cartermill Publishers, London. (Regular updates.)

Lister, C. (1996). *European Union and Environmental Law.* Wiley, London.

Van Leeuwen, C. J. and Hermens, J. L. M. (1995). *Risk Assessment of Chemicals: an Introduction.* Kluwer Academic Publishers, Dordrecht/London.

US legislation with some notes on Canada and the rest of the world

6.1 INTRODUCTION

As with UK and EU legislation, it is possible to identify information-gathering and controlling elements in US chemical legislation. However, there is also important legislation concerned with land contamination and the development of non-regulatory instruments such as use of a toxic substances inventory, trading in permits to pollute and encouragement of pollution prevention by a variety of non-regulatory instruments. We begin, though, with a description of the general policy and legal framework in the USA and end with a brief reference to what is happening in the rest of the world.

6.2 THE FRAMEWORK

Each of the 50 states comprising the federal union theoretically retain their sovereignty; in principle the national government enjoys only those powers granted to it by the Constitution. In practice, though, the national government has extremely broad powers in areas affecting the environment. Moreover, actions by the federal government, within its spheres of competence, are supreme over state laws. Sometimes a state may defer to the federal government to carry out the mandates of an environmental law if for some reason (e.g. lack of resources) it does not want to do it itself. Normally, states are granted authority to carry out or implement a particular law.

The national government has three branches: legislative (composed of Senate and House of Representatives); the judiciary; the executive. The federal executive branch includes the Cabinet departments such as the Department of Justice (DoJ) as well as agencies, such as the Environmental Protection Agency (EPA). The governments of most states are similarly arranged. Local level government may also operate in the environmental sector, but enjoying only the powers delegated to it under state law.

Almost in parallel with the EU, the principal source of powers for environmental protection by the federal government comes from powers vested to it through the Constitution to regulate interstate and foreign commerce (US Constitution Act I, section 8, clause 3). The power is not only concerned with pollutants that may move across state boundaries but with conduct that may be only local but as a result of "like conduct by others similarly situated" may affect foreign or interstate commerce. This commerce clause is unique in that the Constitution vests Congress with virtually unlimited authority to avoid interstate restrictions in commerce. Congress can use its powers to legislate directly; or it can decline to legislate but nevertheless bar states from some or all legislation in the area under a corollary doctrine of the supremacy of federal law known as pre-emption. Other powers which are of importance from an environmental perspective are the power to make treaties with foreign governments and the power to make expenditures to further the public welfare.

Although the commerce clause empowers Congress, the latter can by legislation empower federal administrative agencies to act for the government. The agencies in turn can promulgate regulations; the Administrative Procedures Act 5 sets out rules for such rule making. Here there is a significant difference with Britain where the regulatory authorities cannot make law, only interpret it in the form of guidance notes and, of course, implement it under appropriate legislation (see pp. 57 & 58). The agencies' senior personnel are appointed by the President with the advice and consent of the Senate.

The EPA was assembled from components of various federal agencies, including the Federal Water Quality Administration from the Interior Department, the Pesticides Regulation Division from the Agriculture Department and the Office of Pesticide Research from the Department of Health Education and Welfare. All this was pursuant of an Executive Order signed by President Nixon in 1970.

The EPA implements federal laws through regulations, guidance and interpretation rulings. It establishes environmental standards primarily through administrative rule makings that are preceded by a notice and period during which written comment is allowed. Federal administrative regulations of general applicability are published in the *Code of Federal Regulations* (*CFR*) consisting of 50 titles (as of July 1994): Title 40, Protection of the Environment (Parts 1–799), consists of 16 volumes and pertains to the US EPA and the Council of Environmental Quality (CEQ – which has the lead authority to carry out the provisions of the National Environmental Policy Act (NEPA) – see below).

Other federal agencies that have some environmental responsibilities include: Department of Agriculture; National Oceanographic and Atmospheric Administration; the Army Corps of Engineers; Department of the Interior; Nuclear Regulatory Commission.

There are many US environmental regulatory laws that impact on chemicals. In what follows we concentrate on the main ones concerned with toxic substances, the quality of waters and atmosphere, waste, etc. It should be noted, though, that US laws often have finite lives and come up for reauthorisation regularly so there is some dynamism within the system and what follows should only be taken as a snapshot at the time of writing.

6.2.1   Pollution prevention

Finally, before the systematic review of legislation, it is important to draw attention to an instrument that since its adoption in 1990 has been influencing the general framework of environmental protection, especially with respect to toxic substances. This is the Pollution Prevention Act, that established pollution prevention as a national objective, shifted the focus from end-of-pipe solutions to the sources of the problem and encouraged the EPA to think about using a range of methods, not just command and control instruments to achieve this. The law directs the EPA to establish a hierarchy of approaches to pollution prevention, starting with source reduction and then going from recycling and/or reuse, to treatment and finally, as a last resort, ultimate disposal. As far as the preferred option, source reduction, goes, the law directs the EPA to: facilitate appropriate techniques by business and other federal agencies; establish standard methods of measurement; review existing regulations to determine their contribution; investigate opportunities to use federal procurement to encourage source reduction; develop improved methods for public access to appropriate data; develop a training programme.

6.3   TOXIC SUBSTANCES CONTROL ACT (TSCA)

This was adopted as far back as 1976 to provide powers both to enable collection of relevant information on chemicals and to allow control of their manufacturing, storage, marketing, import, use and disposal. Formally, the Act says in the preamble:

1. Adequate data should be developed with respect to the effect of chemical substances and mixtures on health and the environment and that the development of such data should be the responsibility of those who manufacture and those who process such chemical substances and mixtures (so the polluter pays; though, in general companies are not required to carry out specific testing – see below).
2. Adequate authority should exist to regulate chemical substances and mixtures which present an unreasonable **risk** of injury to health or the environment (so, in principle, the controls are risk based), but
3. Authority over chemical substances and mixtures should be exercised in such a manner as not to impede unduly or create unnecessary economic barriers to technological innovation while fulfilling the primary purposes of this Act to assure that such innovation and commerce in such chemical substances and mixtures do not

present an unreasonable risk of injury to health or the environment (so the measures taken should be balanced against economic impact).

Excluded from the compass of the Act because they are dealt with elsewhere are mixtures, pesticides, tobacco or tobacco products, nuclear matter, food and food additives, drugs and cosmetics, but included are detergents. The Act provides the Administrator of the EPA with general powers, the details of which are specified by rules.

6.3.1   Distinction between new and existing chemicals

Section 8 of the Act requires the Administrator to compile, keep current and publish a list of chemical substances manufactured and processed in the USA prior to 1 July 1979. These are existing chemicals. New chemicals are not on this inventory and are those that are proposed to be manufactured. These are added to the inventory after they are approved for manufacture and industry submits a notice of commencement to the EPA. This inventory now consists of more than 70 000 chemicals (Table 1.1).

6.3.2   New chemicals

The Act (under section 5) requires that manufacturers or importers of new chemicals or manufacturers deemed to be producing a chemical substance for a significant new use must, prior to manufacture (cf. the requirements of 67/548/EEC which is prior to substances being placed on the market; p. 60) notify the EPA. The Agency has 90 days to review the data and decide whether or not it presents unreasonable risks.

The pre-manufacture notice (PMN) must include: the identity of the chemical, production volume, by-products, use, exposure in the workplace, disposal practices, health and environmental effects. All health and environmental data that the submitter possesses must be included, as must any known or reasonably ascertainable data. But unlike the EU programme, no testing is actually required; this partially arises out of the differences in philosophy between a pre-market and pre-manufacture approach. Only $c.$ 5% of PMNs contain any chemical fate or ecotoxicological data so the EPA has developed a predictive approach that relies on QSAR techniques (see p. 27). More than 120 SARs, making predictions about toxicity, persistence and bioaccumulation potential, are currently in use. These are updated and checked from time to time. For example, a comparison of EPA predictions from QSARs and EU measurements on ecotoxicity for new chemicals' base set data for 175 chemicals carried out from 1991 to 1993 showed agreement for 87% of values on fish acute toxicity and 79% for *Daphnia* acute toxicity.

On the basis of predictive analysis of the PMN information, the EPA can restrict or prohibit the production or importation of the chemical – a fate that happened to *c.* 10% of the 20 000 or so new substances notified between 1979 and 1990. Each new chemical that successfully completes the PMN review is added to the inventory of existing substances. There is also a scaled-down review for chemicals manufactured or imported at <1000 kg p.a.; and those chemicals identified as polymers (m.w. (molecular weight)> 1000) are excluded from PMN.

6.3.3   Existing substances

The Act also provides EPA (under sections 4, 6 and 8) with information-gathering and controlling powers for existing chemicals. A testing scheme has been developed by the Office of Pollution Prevention and Toxics – that has the responsibility of regulating existing chemicals. The test guidelines are available (published in 40 *CFR*, Parts 795–797) and used as standards in test rules or consent orders that must be followed. The testing scheme is organised into a four-tiered progression from simple to more complex, less costly to more expensive, and from shorter to longer term. Triggers that are used for shifting between tiers are illustrated in Figure 6.1. The testing required must be specific with respect to the types of effects being evaluated – and costs of tests must also be considered. There are a number of aquatic and terrestrial guidelines that specify acute, chronic and bioaccumulation tests at various tiers that have been published in the Federal Register. Other guidelines have been developed and written for specific chemicals.

In formulating controlling rules the EPA must demonstrate unreasonable risks, and there is a developing ecological risk assessment framework within EPA. However, this is not attached to specific legislation and is neither a procedural guide nor a regulatory requirement. It is intended to further a consistent approach and use of terminology. Under TSCA EPA has, nevertheless, tended to use the most simple, acceptable and valid method to compare PECs with PNECs (see p. 27), but more sophisticated methods that take account of distributions of species sensitivities and exposure concentrations are increasingly being contemplated.

In developing controls EPA must also take into account benefits for various users and the availability of substitutes; the reasonably ascertainable economic conse-quences of the rules, after consideration of the effect on the national economy, small business, technological innovation, the environment and public health. Moreover, it has to be demonstrated that action under TSCA is more appropriate than under any other federal law; for example, air, water, waste and workplace laws.

No wonder that the EPA has made less progress with this element of the Act than with those relating to new substances! To mid-1992 only PCBs, CFCs, the use of certain asbestos in schools and disposal of dioxins were regulated under these powers. However, as part of a revitalisation programme following a critical report by the General Accountancy Office, the EPA has now established a revised risk manage-ment (RM) process. First, chemicals are screened to identify environmental and health risks. These chemicals may come from section 4 activity (see above) and referrals

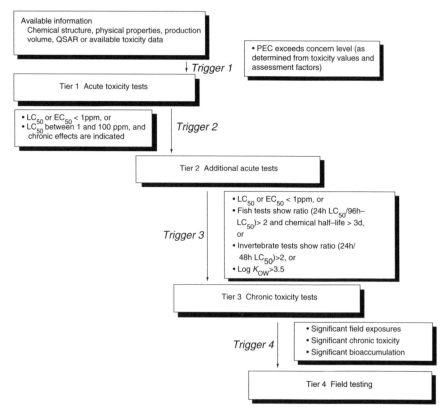

Figure 6.1 Triggers for aquatic tier testing of existing substances under TSCA. After Zeeman, M. & Gilford, J. (1993). Ecological hazard evaluation and risk assessment under EPA's Toxic Substances Control Act (TSCA): an introduction. In W. G. Landis *et al.* (Eds). ASTM, Philadelphia

from other programmes (e.g. OECD) or agencies. These chemicals move into RM1 when initial qualitative conclusions regarding risks are reached. They then move on to RM2 where there are four options: the chemical may be (1) placed on a master testing list for priority testing; (2) placed on a risk reduction list; (3) placed on a regional activities track if problems are likely to be geographically discrete and control will require close coordination with state or local authorities; (4) dropped from further consideration. More than 2000 chemicals have been screened under the programme, of which 400 have gone on to RM1, and of these, 152 were sent on to RM2.

Another significant requirement imposed by TSCA is that chemical manufacturers, processors and distributors must submit reports within 15 working days of learning about substantial risk of injury to health or the environment.

Finally the pollution prevention approach (section 6.2.1) has had an effect on both the new and existing substances' elements of TSCA. As far as new substances are concerned, source reduction plans have been accepted as possible mitigation of concerns for ecotoxicity; and some reduction requirements have been made of some substances should excessive quantities of release of chemicals with possible ecotoxic effects be expected. Pollution prevention requirements can also be written into consent orders for existing chemicals and, of course, pollution prevention is an integral part of the RM programme. The EPA has also proposed a new voluntary programme – called Environmental Leadership – which encourages companies to go beyond regulatory compliance with TSCA to incorporate the prevention philosophy into all their operations from product design and purchasing to manufacturing, marketing and distribution. Another programme in a similar spirit is called Design for the Environment which seeks to encourage and help industry to design, develop and use products in ways that will eliminate or reduce pollution. As far as chemicals are concerned, QSARs are being used to help find substances, within particular categories, with lowest environmental impact.

6.4   THE FEDERAL INSECTICIDE, FUNGICIDE AND RODENTICIDE ACT (FIFRA)

The Federal Insecticide, Fungicide and Rodenticide Act (FIFRA) of 1947 was the original legislation intended to protect the public and the environment from pesticides and designated the Department of Agriculture as the regulatory body. In 1972 FIFRA was comprehensively revised as the Federal Environmental Pesticide Control Act (FEPCA) and designated the EPA as the regulatory body – but FIFRA remains the common acronym.

This legislation requires registration and labelling of any substance deemed to be a pesticide on the basis of intended use, allowing a broad range of authority over chemicals regardless of their original purpose of manufacture. The EPA Administrator is required to publish periodic guidelines and revisions of procedures and data required in support of registration. A manufacturer wishing to make a new pesticide must register it with the EPA. This procedure includes submission of test data, proposed uses and suggested labelling. So this is a hazard/risk-based procedure. Subsequent risk management might include banning, control or restriction of manufacture, use, import or disposal. EPA also has powers to remove from the market products that do not meet its standards of environmental protection, but in doing this it must weigh the impact to cancel against the effects of this on food production and prices. So it is a balancing-based measure.

An adequate label must contain: the name of the product and the manufacturer; the net content of the package; a complete list of all active ingredients; a valid registration number; directions for use which, if complied with, are adequate to protect the public and the environment. The label may also contain warning or cautionary statements. The EPA Administrator is empowered to classify a registered product for general use if it will not cause "unreasonable adverse effects in the environment" or for restricted use if it may cause such effects in the absence of regulatory restrictions. All registrations are automatically cancelled after five years unless the registrant actively pursues registration.

The establishment producing the product must also be registered with the EPA. The application for the registration of an establishment must include data on the types and contents of pesticides currently produced, sold or distributed by that establishment.

The EPA has responsibility under the legislation to set tolerances and action levels for insecticides in food. Here it interacts with the Food and Drug Administration (FDA) whose policies are defined under the Federal Food, Drug and Cosmetics Act (FFDCA). Section 706B FFDCA is known as the Delaney Amendment – after the Congressman who introduced it. It banned from use in foods, drugs and cosmetics any chemical that had been found to induce cancer when ingested by man or animals; in other words it sets out an automatic tolerance of zero for any carcinogen – i.e. there is no safe dose.

6.5   THE CLEAN AIR ACT (CAA)

In its current form, the Clean Air Act (CAA) is largely the result of significant amendments in 1977, 1981, 1990 and 1993. It establishes uniform nationwide standards for new sources as well as for existing sources that have undergone significant modification or change in their operation. It is mainly a controlling law.

The Act vests the EPA with authority for establishing atmospheric standards for particularly hazardous pollutants, giving it the authority not only to designate which pollutants are subject to its provision, but also authority to promulgate emission standards after public hearing and to do this not only on the basis of proof of harm but knowledge of the risk of harm. It also gives it authority to establish national ambient air quality standards (NAAQSs). Two kinds are envisaged: primary, to protect public health; secondary, to protect public welfare in such things as soils, water, crops, vegetation, synthetic materials, animals, wildlife, weather, visibility and climate. Within 270 days of promulgation of one of the NAAQSs each state was to submit an implementation plan to achieve requirements as expeditiously as possible, but in any event within three years for primary protection and a reasonable time for secondary protection. EPA has issued NAAQSs for sulphur oxide, nitrogen oxides, lead, hydrocarbons, particulate matter and CO. Deadlines were not met and much of the subsequent history of the CAA has been concerned with extending deadlines for compliance.

CAA also imposed emission level reduction standards for new motor vehicles; but deadlines for these have also been extended. There is also now a requirement for private companies, federal agencies (including EPA) and the US Postal Service to take measures to reduce the number of vehicles driven to work during peak hours.

The 1970 legislation also required major new pollution sources to seek permits. These are issued only after establishing that standards for controlled releases are being complied with and the source will employ BAT for all air pollutants – but the cost to industry also has to be taken into account. These standards are nationwide to prevent plants from moving to states with less stringent standards. Either states or EPA may be the regulatory body.

The primary focuses of the Act after the 1990 and 1993 amendments have been to clean up air in urban areas, to regulate toxic substances more stringently and to tackle acid rain. In doing this the CAA sets goals and instead of relying on command and control the intention is to allow industries to find their own cost-effective solutions. Provisions directed at acid rain introduce a market-based strategy whereby emission credits – **allowances** – can be traded. This is set up in two phases: the first requires specific reductions from the 111 dirtiest emitters. The second phase, to start in 2000, requires the 2000 or so remaining utilities to cut emissions. Without sufficient allowances to cover their emissions, utilities can be fined on a per ton of output basis.

Another market-based approach is the clean fuel programme. This requires that fleets of heavy- and light-duty trucks purchase an increasing percentage of clean-fuel vehicles. Companies that purchase such vehicles prior to the deadline earn credits that can be sold to companies that cannot meet the targets, so companies that begin upgrading their fleets before deadline will be able to generate revenue to offset environmental expenditure.

6.6 THE CLEAN WATER ACT (CWA)

There was a Federal Water Pollution Control Act adopted in 1948. However, this was replaced in 1972 by the Clean Water Act that took a command and control approach and set ambitious targets: e.g. "that discharge of pollutants into navigable waters be eliminated by 1985" – needless to say, a goal that was not achieved. In addition to statutory standards for water quality, CWA also established a system of effluent controls for point sources. Under the Act the EPA sets so-called guidelines for effluent quality from various industrial operations and municipal sewage treatment plants. These are minimum, technology-based (i.e. for industry, BAT) levels which pay special attention to a list of 125 priority pollutants. Using these, states issue every operation, whether industrial or municipal, with a permit to discharge called a national pollutant discharge elimination system (NPDES) permit. This specifies the technology to be used and the discharge levels and may take into account the quality of the receiving water. Determining what technology a source must install is an elaborate and technical process and has led to much legal wrangling in the courts between the interested parties including the pressure groups.

The 1987 amendments also attempted to address water pollution from non-point sources. A new provision required states to identify waterways where water quality criteria cannot be met without controlling non-point sources; and to develop management plans to control the sources of these pollutants.

The CWA (section 307 (b)) removal credits programme allows those creating point source emissions to reduce the percentage of removal of pollutants in effluents by a factor relating to percentage pollutant in intake. Hence, the point source may not cause further degradation, but is credited in terms of total discharge with a factor relating to existing water quality.

6.7   COMPREHENSIVE ENVIRONMENTAL RESPONSE COMPENSATION AND LIABILITY ACT (CERCLA OR SUPERFUND) 1980 AND THE SUPERFUND AMENDMENTS AND REAUTHORISATION ACT (SARA) 1986

CERCLA was enacted in 1980 to address actual or threatened releases from sites containing hazardous substances. It arose as a direct result of the August 1978 emergency that occurred when chemicals, buried for a long time, were discovered seeping out of the ground and into homes in the Love Canal area of Niagara Falls (New York State). Remedial work at the site and relocation of some 200 families cost the state in excess of $30m.; there was no federal programme to provide assistance. Owners and operators of facilities where hazardous substances and wastes are stored, treated or disposed of were required to notify EPA by 9 June, 1981 of quantities and types of hazardous substances and wastes, and any known suspected or likely releases of these to the environment. EPA used this information to formulate the original National Priority List (NPL), an inventory of contaminated sites requiring clean-up that EPA revises from time to time. Currently NPL has more than 1000 sites requiring immediate remedial attention. Less than 100 sites have been removed from the NPL because they have been cleaned.

CERCLA gives the federal government power to deal with hazardous substances either by removal or remedial action. It established a substantial billion-dollar trust fund. This is intended to pay for action only on an interim basis or in instances where no party can be held responsible. The primary aim of CERCLA is to require parties connected with the contaminated sites or hazardous substances they contain to pay for the cost of clean-up. Cost estimates for clean-up of the 1000 or so waste sites is in the billion-dollar range. It is anticipated that responsible parties rather than the Superfund will bear much of this expense.

CERCLA imposes strict, retroactive liability upon any party that owned the site from which the actual or threatened release is taking place; which generated, transported or disposed of the substances at the site; or which operated the site. Liability is joint and several amongst all such parties. It is no defence that the party's actions may have been lawful – indeed allowed by permits. Responsible parties are not only responsible for remedial and removal costs but also for damage to natural resources. The President may also issue administrative orders to responsible parties obliging them to take protective measures. SARA added an "innocent landowner" defence to deal with problems faced by lending institutions and others.

In addition to the amendments to CERCLA, SARA contains a self-contained law that addresses emergency planning and community right-to-know. Amongst other things this requires companies to:

- Work with community groups to define how various parties should respond to emergencies involving chemicals.
- Notify local groups if an extremely hazardous chemical is released into the environment.
- Submit descriptions of annual releases of chemicals to the EPA for compilation in an annual inventory report (Toxic Release Inventory – TRI). There are more than 300 individual chemicals and 20 categories that must be reported in this way. In line with the Pollution Prevention Act, the TRI, by exposing industrial activities to public scrutiny, has reputedly had a significant effect on pollution avoidance. For example, from 1988 and 1991 there was more than 30% reduction in total releases recorded under the TRI, some of which could be attributed to the direct impact of the TRI.

6.8   OTHER RELEVANT LEGISLATION

6.8.1   Federal Food, Drug and Cosmetic Act (FFDCA)

This is administered by the FDA. It provides powers to assure the safety of foods, drugs, medical devices and cosmetics *for humans*. It contains the famous Delaney Clause (p. 98).

6.8.2   Occupational Safety and Health Act (OSHA)

This is administered by the Occupational Safety and Health Administration. It provides powers for protecting people in the workplace.

6.8.3   Safe Drinking Water Act (SDWA)

This is administered by the EPA. It requires establishment of uniform federal standards for drinking water quality.

6.8.4   Resource Conservation and Recovery Act (RCRA)

This is administered by the EPA. It establishes a system to identify wastes and track their generation, transport and ultimate disposal. It also encourages recycling and alternative energy sources.

6.8.5   Hazardous Materials Transportation Act (HMTA)

This is administered by the Department of Transport (DoT). It gives powers to the Secretary of Transport to regulate transportation of hazardous materials, within and between states and internationally, requiring labelling and appropriate packaging.

6.8.6   Consumer Product Safety Act (CPSA)

This is administered by the Consumer Product Safety Commission (CPSC) and is concerned with consumer protection.

6.8.7   Federal Hazardous Substances Act (FHSA)

This is administered by the CPSC and protects consumers using hazardous substances.

6.8.8   National Environmental Policy Act (NEPA)

This is a wide-ranging piece of legislation that basically requires every federal agency to prepare a detailed statement of environmental impact for each major federal action that may significantly affect the quality of the human environment. This not only includes physical construction but **any action** likely to affect the environment.

Thus the FDA has used the provisions to introduce enforcement requirements for assessing the environmental impact of drugs. A requirement for environmental impact statements (EISs) is the centrepiece of the statute. These provisions have wide-ranging effects on, for example, public and private sponsors, permit and licence applicants, those seeking grants and other financial aid, whether inside or outside the USA. NEPA does not set performance standards, but the culture of requiring some level of disclosure and analysis of impacts of public and private actions requiring federal permits or funding acts to minimise or eliminate impacts.

Many agencies have defined their own regulations detailing internal NEPA procedures defining when EISs are required, how they should be produced (the responsibility of the EPA, not those doing the job) and assessed, and for consultation with other agencies. CEQ has provided regulatory guidance and must also approve implementation plans.

6.8.9   Wildlife and Wilderness Acts

There are a series of Acts that seek to protect nature and therefore can have regulatory implications for chemical substances. These involve: Endangered Species Act; Marine Mammal Protection Act; Wilderness Act; Fish and Wildlife Co-ordinating Act.

6.9   RÉSUMÉ AND COMPARISONS WITH EUROPEAN LEGISLATION

- The basic structure of US legislation is similar to European legislation. In some respects, though, it is more compact, being written into no more than 10 pieces of major legislation, but that is complicated by the fact that states can and do make their own environmental protection laws. Its dependence on specific rule making gives it more flexibility but also in some sense less generality. It also opens up the procedure to more action and hence refinement in the courts which, given the litigious and adversarial system in the USA, can frustrate environmental protection.
- There are five main areas of legislation covering: controls on substances (TSCA and FIFRA); controls on environmental quality (CAA and CWA); controls on current intentional dumping of wastes (RCRA); powers to address previously contaminated land (CERCLA/SARA); controls on the movements of hazardous materials (HMTA). Both TSCA and FIFRA require labelling but not according to a detailed scheme such as required by the equivalent EC legislation (p. 62). Risk management is done on the basis of risk-assessment procedures, but in both TSCA and FIFRA tempered with a need to balance environmental risks from the product against benefits derived from it in economic and social terms.
- Both CWA and CAA define pollution in terms of alterations to the ambient quality – chemical, physical and biological. This could therefore be taken to mean contamination (p. 34); but reference to "integrity" in the CWA suggest that something more is implied. Both recognise the need for the control of especially dangerous substances and both identify "priority lists" (cf. EC legislation; p. 72). Regulation is largely through emission controls (that in CAA may be tradable) based on environmental quality standards (p. 34) and available technology – with some account of cost of implementation being taken into account (so it is BAT with some NEEC). But the Pollution Prevention Act is putting more emphasis on reduction at source.
- RCRA has set up a cradle-to-grave tracking system for waste. CERCLA/SARA in addressing earlier contamination imposes strict liability (p. 36) that is joint and several. This latter also introduces provisions to guard against major accidents and to keep the public informed of potential dangers.
- Finally, HMTA makes provision for controlling substances in transit both nationally and internationally by requiring appropriate documentation.

- Both RCRA and CERCLA are concerned with protection of the soil and in this sense also complement the habitat-orientated provisions of CWA and CAA. This immediately draws attention to a weakness in the US system: the lack of an explicitly integrated approach to all environmental media, something that is being addressed in current EC legislation (p. 87), and that, at least in principle, has been a part of UK legislation ever since the introduction of IPC...in the Environmental Protection Act 1990 (Part I; p. 87).

6.10   CANADA

Canada is introducing a toxic substances management policy (TSMP). It aims at virtual elimination from the environment of substances that derive from human activity and are toxic, persistent and bioaccumulative. These are referred to as Track 1 substances. It also aims at life-cycle management of all other substances of concern. These are referred to as Track 2 substances. In TSMP toxicity criteria are defined under the Canadian Environmental Protection Act (CEPA). The system is therefore hazard rather than risk based with only reference to exposure through persistence and bioaccumulative indicators.

6.11   REST OF THE WORLD

US EPA publishes a *World at a Glance* (*WAAG*) report on international chemicals programmes. The 1995 edition reported that 44 out of 166 countries contacted had some kind of chemical controls programme. These can include provisions both for the collection of information and regulation – but not all either integrate both or indeed carry both out. Few carry out risk-assessment procedures.

6.12   FURTHER READING

Jacoby, D. & Eremich, A. (1991). Environmental liability in the United States of America. In *Environmental Liability* (Ed. P. Thomas), pp. 63–93, Graham & Trotman, London.

Kokoszka, L. C. & Flood, J. W. (1989). *Environmental Management Handbook*. Marcel Dekker, New York.

Smrchek, J., Zeeman, M. & Clements, R. (1995). Ecotoxicology and the assessment of chemicals at the US EPA's office of Pollution Prevention and Toxics: current activities and future needs. In *Making Environmental Science* (Eds J. R. Pratt, N. Bowers & J. R. Staufter), pp. 127–157, Ecoprint, Portland.

US EPA (1995) *World at a Glance: a Directory of International Chemicals Programmes*. Office of Pollution Prevention and Toxics, Washington, DC.

Zeeman, M., Auer, C. M., Clements, R. G., Nabholz, T. V. & Boethling, R. S. (1995). US EPA regulatory perspectives on the use of QSAR for new and existing chemical evaluations. *SAR and QSAR in Environmental Research*, 3, 179–201.

7

International organisations and programmes

7.1 INTRODUCTION

"Chemical risks do not respect national boundaries", a quotation from Chapter 19 of Agenda 21, the policy document that emerged from the second UN Conference on Environment and Development in Rio 1992. The quotation recognises the potential international dimension of chemical releases and their consequences through global dispersal in the atmosphere, in rivers that can often cross national boundaries and in the oceans. It is not surprising, therefore, that a number of international programmes involved with risk assessment and management of chemicals have been established through a variety of international conventions and mechanisms. Some of these establish voluntary agreements, codes of practice, resolutions, declarations and guidelines, whereas others may be legally binding. In developing these, however, there has not always been effective coordination between international and national programmes. This ought to be a major outcome of the Agenda 21 initiatives; intergovernmental meetings have been held in London in December 1991 and Stockholm 1994 to explore this further.

This chapter will review the international programmes. Chief amongst these are the ones deriving from UN activities and it is convenient to begin with those. A position paper – *Current International Activities Concerning Risk Assessment and Risk Management of Chemicals* – prepared for the UN London meeting (1991) referred to above has provided a basis for what follows.

7.2 UN PROGRAMMES

The UN itself was established in 1945 to maintain international peace and security and to achieve international cooperation in solving international problems of an economic, social, cultural or humanitarian character. The promotion of chemical safety programmes is largely undertaken within the framework of the United Nations

Environment Programme (UNEP) and some specialised agencies, especially the International Labour Organisation (ILO), the Food and Agriculture Organisation of the United Nations (FAO), the World Health Organisation (WHO) and the International Maritime Organisation (IMO).

7.2.1 UNEP

This is a programme, not an agency. Its main aim is to continuously review the world environmental situation, to develop and recommend policies on the protection and improvement of the environment, to promote international cooperation at all levels in the field of the environment, and to catalyse environmental work in and outside the UN system. Most of UNEP work on chemicals has been carried out in the context of the International Register of Potentially Toxic Chemicals (IRPTC). This is a component of Earth Watch which was initiated by UNEP to gather and disseminate information on the environment.

The main tasks of IRPTC are to collect and collate data on chemicals, to run a query response service, to issue publications, to implement the London Guidelines for the Exchange of Information on Chemicals in International Trade (see below) and to organise training activities. IRPTC maintains a database and produces a bulletin.

Conventions and protocols that have been developed under the auspices of UNEP are as follows:

- Vienna Convention for the Protection of the Ozone Layer – elaborated and signed in 1985. Signatories shall take appropriate measures in accordance with the provisions of the Convention and of those protocols in force to which they are party to protect human health and the environment against adverse effects resulting or likely to result from human activities which modify or are likely to modify the ozone layer.
- The Montreal Protocol on Substances that deplete the Ozone Layer – adopted in 1987 and amended at subsequent meetings, it provides for the successive phasing out of chlorofluorocarbons (CFCs) and other specified substances likely to deplete the ozone layer.
- The Basel Convention on the Control of Transboundary Movements of Hazardous Wastes – adopted in 1989, it aims to encourage countries to cut back on the quantity and toxicity of the wastes they generate, to manage these in an environmentally sound way, and to dispose of them safely and as near to the source of their generation as possible.
- The London Guidelines for the Exchange of Information on International Trade – initially adopted in 1987, and after two years the concept of prior informed consent (PIC) was incorporated. If export of a chemical that is banned or severely restricted in the state of export occurs, that state shall ensure that necessary steps

are taken to provide the designated national authority in the state of import with relevant information.

7.2.2   ILO

This was initially established in 1919 under the Peace Treaty of Versailles alongside the League of Nations, and became a specialised body of the UN in 1946. Its main purpose is to promote social progress *inter alia* by adequate protection of the life and health of workers in all occupations; so it is largely concerned with the working environment. In this context, two major undertakings are the International Centre on Occupational Safety and Health Information (CIS) and the International Programme for the Improvement of Working Conditions and Environment (PIACT). It has been instrumental in the adoption of various conventions and regulations that have direct or indirect relevance to safety in the use of chemicals at work, and has supplemented these by issuing codes of practice, manuals and technical guides.

7.2.3   FAO

This was established in 1945, its main purpose being to raise levels of nutrition and standards of living of the peoples and to secure improvements in the efficiency of production and distribution of all food and agricultural products. Its involvement in chemical safety is mainly due to the importance of pesticides in agricultural production and the use of chemicals as food additives. An FAO International Code of Conduct on the Distribution and Use of Pesticides established voluntary standards relating to pesticide management, testing, reducing health hazards, distribution, labelling, packaging, storage and disposal.

7.2.4   WHO

This was established in 1946 to promote the attainment by all peoples of the highest possible level of health. Much of its work is concerned with assessment of potentially toxic chemicals, especially within the International Programme on Chemical Safety (IPCS) established in 1980 as a cooperative programme with UNEP and ILO. It assesses information on the relationship between exposure to specific chemicals and harmful effects on humans and the environment, identifies gaps in existing knowledge and indicates research needed to increase understanding, promoting the development, improvement, validation and use of methods for laboratory testing, etc. IPCS now plays a role within Agenda 21 actions (see below).

7.2.5   IMO

This was established in 1978 as the International Maritime Consultation Organisation (IMCO) and was renamed as IMO in 1982. It aims to facilitate cooperation amongst governments on technical matters affecting international shipping, including special responsibilities for safety of life at sea, and for the protection of the marine environment through prevention of pollution at sea. It has been instrumental in the development of a number of legal instruments aimed at encouraging the safe transport of dangerous, hazardous and harmful cargoes at sea. IMO is secretariat of the Convention on the Prevention of Marine Pollution by Dumping Waste and Other Matter, 1972 (the London Dumping Convention), within which comprehensive procedures for the disposal at sea of hazardous and other wastes have been developed.

7.2.6   Other UN organisations

Of relevance are the United Nations Educational, Scientific and Cultural Organisation (UNESCO) established in 1945 in which the Intergovernmental Oceanographic Commission (IOC) and Man and Biosphere Programme (MAB) pay some attention to pollution by chemicals. The United Nations Industrial Development Organisation (UNIDO), established in 1966, encourages and assists developing countries to promote and accelerate their industrialisations. Its provisions include hazardous waste management, or waste reutilisation. Attention is also paid to the safety implications of pesticides. The World Bank, established in 1944, gives loans for large-scale development programmes, but these should take account of environmental impact and the potential for chemical pollution.

Finally, Agenda 21 (see above) makes explicit reference to the management of chemicals in Chapter 19. It is especially concerned with how the use of chemicals can be better sustained by the collection and collation of information, the preservation and enhancement of a technical base with expertise and understanding of risk assessment, and the transfer of these skills and information to the developing world.

To this end the International Conference on Chemical Safety in Stockholm 1994 established the Intergovernmental Forum on Chemical Safety. This is a non-institutional arrangement intended to provide advice and, where appropriate, to make recommendations to governments, international organisations, intergovernmental bodies and NGOs involved in chemical safety on aspects of chemical risk assessment and environmentally sound management of chemicals. One of the aims of the forum will be to provide improved delineation and mutual understanding of rules, initiatives and activities of the major players in chemical safety provisions. It is intended that the forum will meet every three years but with an intersessional group meeting between sessions.

In 1994, the Intergovernmental Forum on Chemical Safety called for an increasing number of industrial chemicals to be subject to internationally accepted assessment.

To this end it has been recommended that IPCS produce Concise International Chemical Assessment Documents (CICADs). These are to provide a precise evaluation of risks to human health and/or the environment resulting from exposure to a particular chemical, including an extensive summary of key studies supporting the evaluation and practical advice for risk management. Most of these would come from extant international and national documents. To be acceptable material will undergo peer review according to agreed criteria. At the time of writing a pilot project involves preparation of CICADs on a small number of substances.

7.3   ORGANISATION FOR ECONOMIC COOPERATION AND DEVELOPMENT (OECD) CHEMICALS PROGRAMME

This was established in 1971 and significantly expanded in 1978. It is a subsidiary body of the OECD Environment Committee. Its main purpose is to assist member countries in risk management and provide a forum for cooperation and, as appropriate, harmonisation of measures, i.e. to avoid unintended barriers to trade. Recommendations and decisions of the OECD are nominal rather than legally binding, though as indicated below, they have influenced the development of EU instruments. Of particular importance are:

- **Principles of good laboratory practice (GLP)** – these address the requirements for testing with regard to facilities, materials, personnel and quality assurance. Methods of receiving, handling, sampling, characterising and storage of test and reference substances are defined; standard operating procedures (SOPs) are required; reporting together with storage, retention and retrieval of records and materials is also prescribed. Again these principles have been influential in the development of GLP philosophy in the EU testing programmes (see p. 60).
- **OECD Guidelines for Testing Chemicals** – that seek standardisation in ecotoxicological procedures. These are influential in the EU test programmes. A good example of how OECD operates is the standardisation of the *Daphnia magna* reproduction study that is intended as an above base-set (level 1) test in the EU programmes (Box 2.1). This test was adopted in guideline form in the early 1980s, but a ring test initiated in 1985, in which two test substances were assessed according to the standard protocol by > 30 laboratories, showed unacceptable variation between laboratories. A workshop of interested parties from regulatory, commercial and academic sectors was organised in 1989 to discuss possible sources of variability. Out of this meeting a general consensus emerged that more fully defined culture systems were desirable and that choice of genetic strain, food and culture would be important. A voluntary research programme designed to identify optimal conditions with respect to these variables was organised and results from this were received in 1991. Clear recommendations were made with

regard to standardisation of genetics, medium and food and a formal updating of the OECD Guidelines was then initiated. But it was agreed that further work was needed to be sure that the recommendations made would lead to reduced variability across laboratories. A detailed report of the 1991 workshop was also sent for comment to all national coordinators of the OECD Test Guidelines Programme and to nominated national experts. In view of the importance and costs of a full ring test of the Guidelines, a pilot ring test was run in 1992 and, following another workshop to consider this in 1993, a full test with three test substances and involving 48 laboratories from 16 OECD member countries and the Czech Republic was initiated with reporting and discussion at a final workshop in 1995. The overall variability between laboratories was much lower than that observed for the 1985 test and, being generally much less than a factor of 10, was deemed by the participants to be acceptable. Again results were sent out to national coordinators and national experts before adoption by the OECD. At all stages, then, there was a considerable amount of involvement by interested parties. In very general terms decisions between options were based upon both sensitivity and convenience. This work on the standardisation spanned over 10 years, involved a large number of scientists over a large number of laboratories across the world and so inevitably was very costly in time, effort and money.

In 1981 the OECD adopted a Decision that states that data generation in the testing of chemicals in accordance with the OECD Test Guidelines and OECD principles of GLP shall be accepted in other member countries for the purposes of assessment and other uses relating to protection of humans and the environment.

The OECD is also coordinating a high production volume (HPV) chemicals programme in which member countries are now investigating, in a coordinated way, the risks associated with chemicals produced at high volumes. Voluntarily, countries gather available data in close cooperation with the chemical industry and prepare a Screening Information Data Set (SIDS) in a standard way. Based on the results, chemicals will be classified as: those of low current concern; those on which further information is required; and candidates for review with a view to possible risk reduction. Again assessments are made on the basis of a comparison of predicted no effects concentrations and predicted environmental concentrations (p. 22). All data collected in this programme are not only used in the OECD analysis but are also transferred to IRPTC for inclusion in its database (see above).

The OECD Council has also decided that member countries should be encouraged to establish or strengthen national programmes aimed at risk reduction. OECD risk-reduction activities may involve: sharing and comparing information; cooperative development of risk-reduction measures for substances that pose more global problems; sharing in the development and coordinating national efforts in order to ensure adequacy and to minimise trade and economic disruption. Pilot actions are being carried out with a focus on a few specific substances. As a result, voluntary commitments from industry have been developed on flame retardants (p. 43) and lead. A review of the programme has led to a distinction being made between concerted international action, that would require OECD countries to agree international actions and cooperative action that would require OECD to back national efforts but would not require internationally coordinated action though it could lead to it. Concerted action should take account of: the precautionary principle; the community right to know; use of market forces; existing instruments; costs and benefits; impacts on trade.

7.4 COUNCIL OF EUROPE

This was established in 1949 to achieve a greater unity between its members (26 European countries) for the purpose of safeguarding and realising the ideals and principles which are their common heritage and facilitating their economic and social progress. Work on chemicals began in 1959 when activities were transferred from the Western European Union. One of the first outputs in 1962 was the publication of the "yellow books" with recommendations on the classification and labelling of some 500 chemicals. This was the basis of the EEC Directive in 67/548 on classification and labelling (p. 59). Work on pesticides resulted in the publication of *Agricultural Chemicals* (now simply known as *Pesticides*) first issued in 1962. These publications are influential in international discussions on requirements for approval of pesticides.

7.5 PARIS COMMISSION (PARCOM)

This is an amalgamation of two earlier Commissions known as the Oslo and Paris Commissions respectively, and is sometimes referred to as OSPARCOM or OSPAR. It is responsible for the implementation of the Paris Convention for the Protection of the Marine Environment for the North-East Atlantic. All countries bordering this area are signatory, and the EU is an active participant.

It prepares controls on specific substances through binding "Decisions" and non-binding "Recommendations". The Convention contains general provisions to reduce pollution from land-based sources and offshore installations and to ban all types of dumping at sea. It recognises the need for monitoring and assessment as a basis for decisions. It interacts with other international bodies (see below).

7.6 NORTH SEA CONFERENCE

This began in 1984 to address marine pollution of the North Sea on an international level. The third conference was held in 1990 and the fourth in 1995. Each produces a Ministerial Declaration applied to the protection of the coastal waters of the nine North Sea states (Belgium, Denmark, Germany, France, the Netherlands, Norway,

Sweden, Switzerland and the UK). The Fourth North Sea Conference agreed a Ministerial Declaration identifying further priorities for action on seven topics. For the prevention of pollution by hazardous substances the Conference agreed a strategy of short- and long-term measures. The Conference Declaration supported the precautionary principle as the guiding principle, and set something of a challenge by adopting a commitment to prevent pollution by:

> continuously reducing discharges, emissions and losses of hazardous substances thereby moving towards the target of their cessation within one generation (25 years) with the ultimate aim of concentrations in the environment near background values for naturally occurring substances and close to zero concentrations for manmade synthetic substances.

Ministers also agreed that scientific risk assessment was a tool (but not neccessarily an essential component) in setting priorities and developing action programmes.

Follow-up action from the Fourth North Sea Conference is largely taken forward internationally by the EU and the Oslo and Paris Commissions. For example OSPAR's annual meeting took place just after the Fourth Conference and considered the commitments arising for OSPAR within their own action plan. Actions arising from the North Sea Conference are taken forward by OSPAR working groups.

7.7   EUROPEAN COMMUNITY/UNION

Measures from the EU have been addressed separately in earlier chapters. It is worth noting here, though, that in its own right it is a party to a number of international conventions and agreements concerning the environment such as PARCOM. It also participates fully in UNEP, OECD and the Council of Europe. In these contexts the EU may speak for all Member States (assume competence) or more usually a consultation is arranged to attempt to define a common EU policy.

7.8   NGOs

Several non-governmental organisations (NGOs) have also been involved with chemical controls. These include:

- European (Chemical Industry) Centre for Ecotoxicology and Toxicology of Chemicals (ECETOC)
- International Council of Chemical Associations (ICCA)
- International Federation of Chemical, Energy and General Workers Union (ICEF)
- International Union of Pure and Applied Chemistry (IUPAC)
- Scientific Committee on Problems of the Environment (SCOPE)

There are also numerous trade organisations representing, for example, the interests of lead, copper, cadmium, detergents, the oil industry and pesticide producers.

Also of considerable importance are the environmental pressure or lobby groups – and well known amongst these are Friends of the Earth (founded in the USA in 1969) and Greenpeace (founded in Canada in 1972), both of which take an active interest in chemicals and their impact on the environment. The World Wide Fund for Nature (WWF), founded as the World Wildlife Fund in 1961, takes an interest in chemical controls and pollution. Note should also be made of the European Environmental Bureau (EEB), the objectives of which are to bring together environmental NGOs in the Member States of the EU to strengthen their effect and impact on EU environmental policy and projects. A key activity of all these organisations, either explicitly or implicitly, is to obtain information on the state of the environment and on the performance of the major players in order to use public opinion and pressure to enhance environmental protection through their influence as an electorate and as consumers.

7.9   STANDARDS ORGANISATIONS

There is a large number of standards organisations and their work covers a wide range of subjects. It is relevant in environmental protection and chemical control in that it ensures standardisation in methods and methodology and hence facilitates the mutual acceptance of information and classifications between interested parties on an international, national and local scale.

The major players can be organised in a geographical hierarchy from international, involving OECD (see above) and the International Standards Organisation (ISO), to regional, involving the Comité Européen de Normalisation (CEN) in Europe and the American Society for Testing and Materials (ASTM) in the USA and, finally, to a national level. A list of some of the latter is given in Box 7.1.

Involvement of these various bodies has included standardisation of:

- Environmental measurement methods
- Measurement methods for environmental properties of chemical substances and chemical products
- Pollution control methods and equipment
- Environmental management tools
- Methods for the evaluation of the environmental effects of products

We have already seen how the standardisation of general approaches to ecotoxicity testing (GLP, p. 60) and of specific tests (largely by OECD and ASTM) have had an important part to play in the gathering of information, the assessment of risks and the prioritising and classification of chemicals. It is often argued that this is as far as these organisations go in risk management; legal bodies set levels and standards

Box 7.1   Standards institutions

AFNOR (France)
Association française de normalisation
ANSI (USA)
American National Standards Institute
BSI (United Kingdom)
British Standards Institute
CEI (Italy)
Comitato elettrotecnico italiano
DIN (Germany)
Deutsches Institut für Normung e.V.
DS (Denmark)
Dansk Standardiseringrad
ELOT (Greece)
Hellenic Organisation for Standardisation
IBN (Belgium)
Institut belge de normalisation
(Belgisch Instituut voor Normalisatic)
IIRS (Ireland)
Institute for Industrial Research and Standards
Luxembourg
Inspection du travail et des mines
NNI (Netherlands)
Nederlands Normalisatie Instituut
UNI (Italy)
Ente nazionale italiano di unificazione

organisations ensure that their achievement is by mutually acceptable, prescribed procedures. However, the distinction is not necessarily as sharp as this. By playing a part, for example in specifying methods for evaluating products and processes and providing a label to indicate that this has been achieved and verified (e.g. BS 7750, p. 40) they provide information in the market-place and hence influence market forces (see p. 38). Certainly some thought has been given to how they might be used more broadly in controlling chemicals and this will be worth keeping under review.

The way these standards organisations function is well exemplified by ASTM. Organised in 1898 it has grown into one of the largest standards development systems in the world. It is a voluntary, not-for-profit organisation that provides a forum for producers, users and those having a general interest from government and academia. The standards are written by volunteer members who serve on technical committees. Through a formal balloting process, all members of ASTM can have input before standards are published. Task groups prepare draft standards which are reviewed in subcommittee and committee. Once approved at committee level the document is submitted for balloting to the Society as a whole. All negative votes cast during the balloting process, which must include documentation, must be fully considered before the document can proceed to the next level. The number of voting producers on a committee cannot exceed the number of others. The standards so produced are developed voluntarily and used voluntarily. They become legally binding only when a government makes them so, or when they are cited in a contract.

With a number of organisations with similar remits operating at different levels there is potential for overlap and confusion. Some attempts have been made to minimise this between CEN and ISO through the signing of the Vienna Agreement. Work on standards will be initiated by CEN when:

- A field is not adequately covered by ISO
- The field is a priority area as a result of EC legislation

However, it will avoid starting work when:

- There would be an obvious duplication of effort
- The subject is already dealt with in sufficient detail in EC legislation
- The EU has already made arrangements with other organisations to carry out the work as is the case for test methods for toxicity and ecotoxicity

7.10 RÉSUMÉ

- Pollution control has an international dimension, because many pollutants do not respect national boundaries. So there has to be international activity in the area.
- There are a large number of players in this international arena ranging from inter/ supra-governmental organisations to non-governmental ones.
- The different organisations operate through a variety of instruments ranging from treaties and resolutions with if not "legal force" certainly the force of national honour, to voluntary agreements, guidelines and standards, etc.
- This diversity is not only a reflection of the diversity of problems, but also of the diversity of interest groups and the somewhat uncoordinated way that governments and other bodies operate at an international level.
- The UN ought to be a coordinating force. And the Intergovernmental Forum on Chemical Safety that emerged out of the UN Conference on Environment and Development in Rio 1992 should be a catalyst for this.

7.11 FURTHER READING

Brenton, T. (1994). *The Greening of Machiavelli*. Earthscan Publishers, London.

8

The future

8.1 THE CURRENT POSITION

As will have become apparent from the foregoing, methods for protecting the environment against possible harmful effects from chemicals have evolved in both Europe and the USA towards more risk-based approaches; so the legislation has required both information-gathering and controlling components. The risk-assessment procedures are now more formally defined in Europe. The results of risk assessment are to establish critical levels for point source emissions or to control production, storage, marketing, use and disposal so that critical levels are not likely to be exceeded and this might go as far as complete prohibition.

The emphases have been on the following:

- Chemicals are being dealt with on an individual basis using risk quotient analysis.
- Command and control has dominated as a regulatory instrument.
- The controls have separated manufacturing processes and their emissions from products, including their use and disposal.
- Controls have often been applied in isolation from a consideration of social benefits.

This final chapter looks into that "crystal ball" to consider where current trends might be leading us. First, though, the next section speculates on how successful the controls have been to date and from that on future trends.

8.2 STATE OF THE ENVIRONMENT AND RÉSUMÉ OF TRENDS

The ultimate arbiter of the effect of environmental protection legislation ought to be the state of the environment. But for various reasons, some of which were addressed in Chapter 2, assessing the state of the environment is not straightforward and, even if it were, linking changes, positive or negative, with specific items of policy and legislation would be difficult because of the many factors affecting particular conditions and because of the complex time lags between causes and effects. Thus even if the Montreal Protocol, with regard to ozone-depleting substances (Chapter 7), were fully implemented by all signatories immediately, the current abundance of stratospheric chlorine is estimated to continue to increase over the next several years, peaking around the turn of the century and only reducing to 1970 levels (when the Antarctic ozone layer first appeared) some time towards the middle of the next century.

Reports on states of the environment do nevertheless exist at various geographical levels (Box 8.1), and long-term systematic monitoring is likely to be of increasing interest as a check on particular aspects of the environment. In the EU this will be a

Box 8.1   Environmental audits and assessments

States of the environment

There is a whole hierarchy of reports on states of the environment and some of the main ones are given below. This goes from global to national, but could also include local states of the environment reports from local authorities and reports on the impact of businesses on local environment.

World

- *Environmental Data Report 1991/2*, The United Nations Environment Programme, Basil Blackwell, Oxford (3rd Edition).
- *The World Environment 1972–1992, Two Decades of Challenge*, edited by Mostafa K. Tolba and Osama A. El-Kholy, published by Chapman and Hall on behalf of the United Nations Environment Programme, 1992, 884 pp.

OECD

- *The State of the Environment 1991*, OECD, Publication Service, Paris.

EU

- *The State of the Environment in the European Community*, 1992 (Accompanying the Fifth Action Programme: *A New Strategy for the Environment and Sustainable Development*).
- *Europe's Environment*, The Dobris Assessment, 1995, EEA, Copenhagen.
- *Environment in the European Union 1995*, Report for the Review of the Fifth Environmental Action Programme, 1995, EEA, Copenhagen.

National

- *The UK Environment*, 1992, HMSO.
- *The Environment of England and Wales. A Snapshot*, 1996, Environment Agency.

main responsibility for the European Environment Agency (EEA; Chapter 4); and a major start for this is the assessment of Europe's environment that was initiated out of a Pan-European Conference on the Environment which took place at Dobris Castle (Prague) in 1991. This more than 600-page compendium of data, analysis and comments will form a baseline for future EEA activities, and was a basis for the review it undertook of the progress in the Fifth Action Programme (p. 53).

One interpretation from all these surveys is that "some progress" has been recognised in effectively controlling the traditional sources of pollution from point emissions at least in the developed world; but there are continuing concerns about more distributed sources of pollution, whether they be from use of chemical products in agriculture, transport and households, or from their disposal. The impact of these non-point sources of contamination on the environment through their long-term chronic and possibly subtle effects on ecological diversity and processes is not well understood (e.g. see Box 8.2).

This will inevitably direct attention in the future towards a consideration of:

1. The ability of the science to address and predict the more subtle effects.
2. The need to see production, use of products and their disposal in a more integrated way in addressing the challenge of distributed sources.
3. The need to optimise the use of controlling instruments in addressing the more subtle and distributed causes of pollution.

Another area of continuing concern, though, is with respect to accidents, major and not so major, at production plants and in transit either by land or water. There is a need, encouraged by public concerns, to develop appropriate risk-assessment and management tools to address these accidental happenings more effectively.

At the same time there has been increasing attention given to the balance between benefits to environmental quality gained from controls, and the loss of benefit from the reduced availability and use of chemicals that goes with it. This has appeared as a feature in European, US and British legislation. Thus Regulation 793/93, on existing substances (section 5.2.4), specifies that where the rapporteurs make recommendations for restrictions on the marketing and use of a substance, they must submit an analysis of the advantages and drawbacks of the substance and of the availability of substitutes, so focusing on the benefits (advantages) of the substance in contrast to the risks (drawbacks) it poses for human health and the environment. An Executive Order (12291) from President Reagan (replaced by 12866 issued by President Clinton in 1993) and the wording of TSCA require that the US EPA use cost–benefit analysis as one of its tests in taking major decisions. Clause 39 of the UK Environment Act 1995 places the Environment Agency under a duty in considering whether or in what manner to exercise any power, to take into account the likely costs and benefits of the exercise or non-exercise of the power.

Reasons for this cost–benefit, balancing orientation, are complex; but as indicated in Chapter 1 they include an explicit recognition of the balance between economic

Box 8.2   Endocrine disruptors – an *ecological* challenge for the future?

There is evidence that some chemicals are impairing reproductive performance in humans and wildlife. Many chemicals have the potential to disrupt the endocrine control of reproduction; e.g. simulate female hormones (be oestrogenic) and thereby cause reduced sperm counts in males exposed to them. Substances implicated include organochlorines (e.g. pesticides), alkylphenolic compounds (used in cleaning products), PCBs, dioxins, by-products of the contraceptive pill, and also natural plant and fungal sterols. As is clear, these chemicals derive from many and varied sources, may be the products of secondary reactions in the environment, possibly have their effects in consequence of their association with other chemicals in complex mixtures, and appear capable of causing problems at very low doses/concentrations. They are, therefore, good examples of the more subtle and unexpected effects referred to in the text. They raise important questions, some of which are currently the subject of active research, for example:

- How does the disruption work? And is it universal in the sense that one and the same molecules can disrupt different groups (e.g. vertebrates **and** invertebrates)?
- To what extent are individuals exposed to potential disruptors actually affected by them in complex environments, where, for example, there may be a wide range of natural disruptors from plants and fungi?
- Even if the potential is realised in some individuals, will it matter for populations and ecosystems? There is certainly evidence that tributyltin, used in marine antifouling paints, is capable of disrupting the reproductive system of female molluscs and has caused population decline near harbours.
- To what extent do routine ecotoxicological tests pick up or miss endocrine disruptors? Chronic tests usually do measure reproductive performances, but not necessarily over long enough exposure times or with enough realism in terms of the complexity of exposure.

A couple of interesting references that treat the subject somewhat differently are:

Colborn, T., Dumanoski, D. and Myers, J.P. (1996). *Our Stolen Future*. Little Brown & Company. Boston or London.

Institute for Environment and Health (1996). *Environmental Oestrogens: Consequences to Human Health and Wildlife*. IEH, Leicester, UK.

activity and environmental protection through the sustainable development debate, and a realisation that through the polluter pays principle, costs to the economy from environmental protection are ultimately borne by the consumer.

There follow brief commentaries on each of these trends, starting with the challenges for the science and then going on, in turn, to consider methods of integration, risk assessment of accidental happenings, the introduction of cost–benefit analysis and finally the selection of controlling instruments.

8.3   SOME CHALLENGES FOR THE SCIENCE

There is a sense in which, provided a problem is clearly defined, science can be applied to an attempted solution by a critical analysis involving carefully controlled observation. The problem – what to protect – has both a scientific content, in defining natural states, and a socio-political one, in defining what it is about natural states that society values. So defining which ecological systems are to be protected and to what extent ought to involve discussions between the scientific community and society at large. But this is not easy, often because the concepts and language used by the scientific community are not easily grasped by the non-technical public. There are challenges here on both sides of the divide.

The other important general point is that using science to protect the environment means applying science. This is especially so in applying science through ecotoxicology in risk assessment. This has some important implications for the way the information generated is interpreted. Thus in basic science, scientists are often worried about making false claims, committing Type 1 errors (Chapter 2); and to avoid this, by convention, they set stringent standards for the chances of false positives (claiming differences between control and experimental systems when they do not exist) of one in 20 or even one in a 100. Yet in ecotoxicology it may be more important to avoid false negatives or Type 2 errors (Chapter 2), i.e. of claiming no differences between a control and treatment when they exist, otherwise NOECs, for example, might be set too high. The conventions here are not as clearly defined as they are for avoiding Type 1 errors and require more consideration.

So with these general points in mind, what kinds of ecological targets can we expect to be emphasised in the development of risk assessment and what implications will they have for the developing science? One possible, and in many respects reasonable, starting point is to consider the protection of ecological systems and their processes in terms of the services they provide to humanity; e.g. in terms of the composition of the atmosphere, the quality of soils and water, the availability of resources such as fish as food and lumber as raw materials. The challenges for the science are then to make explicit the links between the ecosystem structures and processes that were described in Chapter 2 and these services – something about which we currently have only a rudimentary grasp – and to devise ecotoxicological test systems and measures for monitoring and surveys that reflect these links. Whether this means more emphasis on processes rather than properties of ecosystems, or on multi-species systems rather than single-species systems seems as yet unclear. But it will almost certainly mean the development of systems that pay more attention to soil ecology and atmospheric processes (cf. current emphasis on aquatic systems), and which pay special attention to defining appropriate endpoints not only from a biological but also a statistical point of view.

The results of these tests and their incorporation into risk assessment always carry some uncertainty, if only because of the complexity of the systems that are being addressed. Surprises will undoubtedly continue to emerge. But in order to be able to recognise problems and respond effectively it will be necessary to have more and more effective monitoring programmes. Again these will be strengthened by our understanding of what should be measured, but also by advances in technology that facilitate remote sensing, automatic data processing, interpretation and presentation of data, etc.

In all this, there is also a general recognition of the need for a flow of information, but most importantly advice and help through training and educational programmes, from the scientific and technical community to society at large and from developed to developing world. This is part of the commitment in Agenda 21 (section 7.2). And given that many of the problems for environmental protection are likely to be emphasised in the developing world, then facilitating this is going to be of crucial importance.

8.4   A BASIS FOR INTEGRATION

Recognising a need for integration in pollution control comes largely from the realisation that removing polluting substances from one waste stream can mean that they simply turn up in others (section 5.5). But it touches on a variety of related issues, and also raises a number of the challenges for science. Essentially there is a need to consider integration across substances (how are they likely to interact as mixtures in the environment?), across environmental compartments (how are we to weigh their effects upon atmosphere, soil and water?), and across the life-cycle of substances (how are we to weigh the effects from production, through use to disposal?).

Formally this can be done by integrating risk assessments across substances (but for an alternative approach, see Box 8.3), media or life-cycle components; e.g. by comparing predicted environmental concentrations with predicted no effect concentrations for each of these elements. The lower these ratios the less likely an effect; but whether this will turn out to be a simple linear relationship seems unlikely and this is something that requires more scientific investigation. These risk quotients can then, in principle, be combined across substances released into particular media in a variety of ways – but the easiest is by addition which assumes no synergy of effect in combination (either negative or positive). There is some evidence that this is a common situation, but more scientific work is needed to investigate the generality of this. The consequences of releases into different media can be assessed by repeating this for each and again combining. But there are problems here: How can risks in, say, aquatic ecosystems be equated with those in terrestrial? Some would say this is not possible: "it's like comparing apples and oranges". Others would argue that it is possible by taking into account the relative sensitivities or importance attached to the different media, for example, by applying appropriate weighting factors. Finally this can all be repeated for different aspects of the life-cycle of a product, taking into account the releases that might be associated with production, use and disposal, again possibly with appropriate weighting factors.

Box 8.3   Direct toxicity assessment of complex mixtures

Many effluents that arise from industrial processes consist of complex mixtures of chemicals. Their effects might be estimated by the integration of assessments of risk as determined in isolation (see text). However, this can be criticised as a somewhat theoretical and reductionist approach. An alternative is to carry out tests on samples of the effluent, diluted in such a way as to take into account conditions in the receiving environment (e.g. flow rates), and the physical extent of the environment that is of concern (i.e. how near to the outflow protection is to be afforded). Controls might then be applied in terms of requiring no effect on prescribed test systems at particular dilutions of the effluent. This methodology has been deployed in the USA and at the time of writing is being considered for use by the Environment Agencies in Britain.

Its advantages are that it is a pragmatic and holistic approach, needing to make no assumptions about how chemicals interact, because it measures effects directly. However, there are still some potential problems: for example, in terms of the relevance of tests with respect to the ecosystems that are at risk, the robustness of the tests in terms of routine use, and the representativeness of the effluent samples that might conceivably vary considerably over both the long- and the short-term. And, there are the usual questions about cost effectiveness, especially given the amount of skill and effort often involved in carrying out ecotoxicological work.

Now in principle this could be repeated for various systems and combinations of production, use and disposal. That option that has lowest impact judged in these terms might be defined as the best environmental option (BEO). Another consideration, though, is the cost of an option. It is very likely that the return in terms of reducing the impact reduces as more and more investment is made in improving production processes, encouraging safer use and finding better ways of recycling or reusing or destroying the discarded product. The option that maximises reduction of impact in the most cost-effective way might be referred to as the best practicable environmental option (BPEO). These are issues that will loom large in future risk reduction programmes. They are certainly issues being considered by the Environment Agency for England and Wales, as part of its integrated pollution control programme.

8.5   ASSESSING AND MANAGING RISKS FROM MAJOR ACCIDENTS

Not much more can be written on this over and above that contained in section 2.6. It is worth reiterating, though, that there are two components involved in assessing risks to the environment from accidents: (1) the likelihood of accidents occurring (which is related to technology and techniques) and (2) the likely consequences for human health and the environment. This latter element is the basis of risk-assessment procedures already described (see section 2.4). The first element, likelihood of release and hence exposure, requires information on likely systems failure as a result of human error or breakdown/malfunction in the machinery; in turn, challenges for social science and engineering.

For management purposes it is important to have these elements characterised as far as possible under normal, abnormal and emergency situations with contingency programmes worked out in the event of an accident.

8.6 TAKING BENEFITS AS WELL AS COSTS INTO ACCOUNT

The point has been made repeatedly that the reason why we produce so many chemicals in such large quantities is that they bring benefits in terms of health, food production and quality of life. The chemical industry is meeting a demand. Yet some chemicals carry risks to human health and the environment so controls are needed. To base these exclusively on a precautionary approach, with the aim of zero emissions, would be both unrealistic and unreasonable. Methods are needed to weigh dispassionately advantages with disadvantages associated with chemicals and the products into which they are incorporated. Of course, with so many chemicals and, especially with organic substances, so many possible permutations in their synthesis, there will usually be substitutes for any function. However, care has to be exercised in using substitutes because they may not be as cost-effective, and often they may be less well studied than the more commonly used substances. So in weighing advantages and drawbacks, the availability of substitutes together with their relative merits ought to be taken into account.

In making decisions about if and how to control chemical substances these considerations about balance of risks, costs and benefits are usually implicit. However, there are now attempts, through what have been described as the balancing approaches in Chapter 1, to make the process more explicit, more transparent and more standardised across regulatory authorities. This is the thrust of EU legislation on existing substances already referred to above.

One approach to this is through an extension of the rationale described in section 8.4: to cost the effects of a particular option for control and to judge the returns in terms of indicators of likely impact on the environment. The costs as originally described in section 8.4 were restricted to the costs of implementing controls for industry through extra management, investment in plant, monitoring, etc., i.e. the cost-effectiveness of the techniques and technologies. But to these costs should be added others: to society in terms of lost benefits arising out of restrictions carried out by the proposed controls; of any lost employment, for example as a result of possible banning of, or reductions in, production of a substance; of any lost competitiveness in the international market-place caused by the economic burdens of the controls. The sacrifices in social benefit can be assessed using standard techniques in welfare economics that amount to valuing benefits in terms of consumers' willingness to pay, taking due account of the effects of supply and demand on prices. When there are a range of options, e.g. of substitute substances and/or processes for production, use and disposal and/or methods of control, it becomes possible at least in principle *to compare* gains from reduced environmental impact as measured by the risk-assessment procedures already described with overall costs to society.

The measurement of impact creates some challenges, however. The comparative approach is appropriate if all circumstances of impact are indeed comparable. It becomes more problematic if the indices of impact are being compared across different environmental compartments or across different parts of the life-cycle of a substance/product. In these situations judgements have to be made about the relative importance of an impact in water as compared with soil or atmosphere and an impact coming from production or use, or disposal options. Sometimes also the comparative approach may not be applicable because only one substance or option is available. Under these circumstances some more absolute judgement is needed on the gains from environmental protection against social costs.

In these more difficult circumstances it is necessary ideally to consider both costs to society and benefits to environment in common value units — so that their relative merits can be judged in exactly the same terms. The obvious units are financial. But the idea is not to put a monetary value on the intrinsic worth of environmental entities; only to use financial units as quantitative indicators of preferences within society where resources are necessarily limited. Similar problems are addressed in putting a financial value on human life so that risks associated with certain activities can be quantified relative to the gains from those activities.

There is now a developing field of environmental economics that is formalising a philosophy and methodology to address these kinds of issues. Fundamentally, as with the valuation of social benefits discussed above, it is an attempt to judge preferences in terms of willingness to pay, either as assessed in real market-places (e.g. commodities such as fish and lumber), or surrogate market-places (e.g. by considering how much extra consumers are willing to pay for accommodation in an environment with "clean air"; or "peace and quiet") or even imaginary market-places (by sounding people out on willingness to pay for a clean environment that they can visit or simply enjoy being aware that it exists). Alternatively it is possible to explore the likely compensation that people claim they would need to put up with a degraded environment. And also to take account of the costs to society of environmental degradation in terms of aversion (e.g. the costs of more insulation to avoid noise) or remediation.

Clearly these valuation techniques are not without difficulties and controversy. But they can be made explicit and transparent, which means they are amenable to scrutiny, discussion and debate. They can, in principle, allow the elements of the cost–benefit analysis to be compared in the same units. Again this comparison is not without further difficulties especially since costs and benefits often fall on different sectors of society and/or on different countries. The latter is especially apposite in an EU context since the economic benefits of production may be enjoyed in one Member State, with the environmental costs being suffered in other Member States. The cost–benefit, balancing analysis neither creates nor solves these problems. Judgements have to be made on a political basis; but cost–benefit analysis can make the considerations more obvious and can, for example, draw attention to social inequity and the need for systems of compensation.

Another problem that applies to both cost and benefit sides of the analysis is that they will normally be spread over many years – which means that allowances have to be made to reflect the cost of capital and, more philosophically, the general social preference for instantaneous benefits rather than putting them off. These allowances are made as discounts. This rationale has to be offset by sustainability considerations that take account of responsibilities for future generations. What that means is that in the context of discounting future values from protecting environmental/ecological entities lower discounts have to be applied than are normally used by economists. However, one would not expect zero discounts to be appropriate here because it would be rare for anyone to place a lower value on their own welfare than they do on that of descendants.

These are the basic principles in the balancing of approaches. How are they put into practice in EU and US legislation? The application of cost–benefit analysis in the balance of advantages and drawbacks of existing substances is in its infancy in the EU. The extent to which authorities pursue the appraisal of the balance of advantages of controlling options is likely to depend on the information available, the proportional relationship between advantages and drawbacks and the scale of drawbacks. It will range from a systematic qualitative approach describing advantages and drawbacks, to a semiquantified one in which welfare benefits are quantified, but environmental returns, though described, are not valued, to a fully quantified one in which not only social benefits but environmental problems that are avoided are valued.

It seems clear that in this procedure the fully quantified approach will not be carried out routinely. But there could be considerable insight from the semiquantified approach, as to likely financial sacrifices for a particular action, especially if a comparative approach of the kind described above could be effected. Even if this were not possible, statements such as banning use of TBT (tributyltin) in antifouling paints for marine shipping could, as seems likely, cost the UK $> £10$ m. p.a. with one-off conversion costs of $> £9$ m., put the controls into an economic context. Furthermore a decision to ban the use of TBT in this example implies that the value of environmental damage so avoided would be in excess of these sums even though currently it has not been possible to estimate these environmental values.

As far as US legislation goes there is, as already noted above, a requirement for cost–benefit analysis associated with TSCA, and a systematic and quantitative methodology has been developed to assess the effects of regulations. The first stage in the process is the preparation of a uses and substitutes (U/S) report, presenting information on how much of a substance is produced and exported, the type and extent of usage, and the advantages and disadvantages of any substitute. The next stage is an economic analysis to identify different regulatory options and to develop fully documented estimations of costs and benefits. This enables policymakers to decide which if any of the regulatory options should be taken. Once a decision is made a regulatory impact assessment

(RIA) is prepared. In principle all benefits of regulatory options should be considered, including environmental/ecological ones; but in practice the emphasis is on costed benefits to human health and avoidance of death.

8.7   RISK MANAGEMENT OPTIONS

The dominant method of control of hazardous substances in the past has been through prescriptive legislation. The use of negative labelling has also played an important part. There are concerns, though, that such measures miss out on the possibility of more innovative approaches that industry would use if there were more flexibility, that by focusing on end-of-pipe they do not take enough account of the need for integration and that this puts too much emphasis on risks and not enough on balancing risks with benefits. These are some of the reasons that are encouraging more consideration of the possibility of using market instruments and voluntary programmes (Chapter 3).

The counter-arguments are that without prescription it will be more difficult to protect the environment because there are likely to be inconsistencies and cheating. The system of production, marketing, use, disposal, etc. is so complex that integration could only be achieved by command and control. It is unlikely to be possible to design market instruments (labels, taxes, etc.) sufficiently effectively to achieve targets through these in terms of environmental quality standards.

There are uncertainties on both sides. What the debate means, though, is that we shall undoubtedly see more attempts at the development of market instruments and more pressure for consolidation of existing instruments so that, for example, those addressing different aspects of the water cycle or waste collection and processing streams are all obviously compatible. Moreover, the use of mixes of instruments in achieving desired goals is likely to become more prominent.

In all this, however, there are likely to be a number of guiding principles as far as hazardous substances are concerned:

1. There are critical concentrations (call them CLs) that we shall want to ensure are not exceeded in the various environmental compartments. How seriously these are taken will depend upon:
   (a) the likely effects at concentrations beyond these;
   (b) the likely generality of effects at concentrations beyond these (*re* species, ecosystem processes);
   (c) the uncertainty about (a) and (b);
   (d) the levels themselves; substances causing effects at parts per million levels will be taken more seriously than those causing effects at more than parts per thousand.
2. On the basis of these considerations it might be necessary to apply some precaution in defining CLs, e.g. by introducing bigger safety margins.

3. On the other hand it is likely that for most substances we can define non-zero levels below which we should not be concerned (call these levels of no concern – LNCs).

4. Between CLs and LNCs there are opportunities for taking benefits of chemicals and costs of controls increasingly seriously, for considering the need for adjusting levels to optimise protection across environmental compartments and for allowing more flexibility in achieving controls.

5. CLs are likely to be prescribed as maximum environmental quality standards, for example by legislation in terms of "clean air", "clean water" and "clean soil". In Europe it will be likely that these are prescribed on an EU basis with the opportunity for more flexibility at a Member State level for measures between CL and LNC – on the presumption that this will not obstruct trade.

6. How they, and indeed any levels between CL and LNC should be achieved in terms of how and when controls will be applied will be importantly guided by risk assessments:

    (a) if they suggest that manufacturing is likely to be the most important source, end-of-pipe restrictions may be appropriate;

    (b) on the other hand if use is a more important source, restrictions may either be in terms of prohibition or use of market instruments, depending on the likely relationship between PECs and CLs or LNCs;

    (c) if waste disposal is the most important source, specification of disposal methods, or influence through use of differential levies, etc. may be more appropriate;

    (d) if the analyses point to local problems specific controls might be applied to particular manufacturers; but if problems are more regional or global, collective action will be needed, possibly not just by a manufacturing sector, but by a community of manufacturers/users that is likely to influence the environment collectively. The extent to which prescription, market instrument or voluntary action will be appropriate is likely again to depend on the relationship between PECs, CLs and LNCs.

7. New substances and products will require special attention. This is in a sense the front line in environmental protection, i.e. with the possibility of preventing problems before they occur. It seems very likely, therefore, that risk assessment will be applied more stringently across the life-cycle of these substances with more emphasis given to precaution and less to benefits that might never be enjoyed!

A last word should be given to the public perception of risks. Experience in the UK, for example with the disposal of the Brent Spar offshore oil installation by Shell and the BSE (bovine spongiform encephalopathy)/CJD (Creutzfeldt–Jakob disease) controversy, indicates that the perception of risks, as communicated and influenced by the media, can make more impression on decision-makers than scientific risk assessments and cost–benefit analyses. Understanding how these perceptions are developed, and are influenced by, and in turn influence, regulators, policymakers and decision-makers will no doubt be an important area of enquiry for the social sciences. Risk perception is clearly crucial in considerations about risk management from both producer and regulatory perspectives.

8.8   RÉSUMÉ

- Chemical control legislation to date has been largely focused on particular substances in particular media and has been prescriptive. It has separated production from use and ultimate disposal and has often ignored benefits derived from substances.
- It is hard to judge how effective this has been – but there are still profound environmental problems, most of which can be linked more or less directly to chemicals. Special concerns are in terms of pollution through non-point sources.
- More emphasis is therefore likely to be focused on pollution prevention at source as emphasised in the US legislation and the EU Fifth Environmental Action Programme. And here voluntary programmes and market instruments may prove more attractive and effective than the traditional prescriptive approaches.
- An integrated approach with respect to substances and environmental media is also likely to become more important and this will present some challenges for the scientific basis of risk assessment.
- Benefits from substances being controlled are being taken more seriously and this raises challenges for carrying out risk/cost–benefit analysis in cost-effective ways.
- The basis of risk perception by the public will undoubtedly become ever more important.

8.9   FURTHER READING

Nature (1996). Briefing – Risk: a suitable case for analysis? *Nature*, **380**, 10–14.
New Scientist (1996). Give us the beef about beef. *New Scientist*, **149**, 3. (Editorial.)
O'Riordan, T. (Ed.) (1995). *Environmental Science for Environmental Management*. Longman Scientific & Technical, London.
UK DoE (1993). *Making Markets Work for the Environment*. HMSO, London.
UK Government/Industry Working Group (1994). *Risk-benefit Analysis of Existing Substances*. UK DoE.

Index

Accidents
  consequence 30
  major hazards 84-5
  probability of occurrence 30
  risk assessment 30–1, 123–4
  risk management 123–4
Acid gases 80
Acid rain 17, 99
Acidification 17
Acute tests 20
Air quality management areas (AQMAs) 80
Air quality standards (AQSs) 80
Alkali Acts 5
American Society for Testing and Materials (ASTM) 113, 114
Animals 50
Aquatic systems 63–4
  pollution controls 72–8
Aquatic tier testing 96
Article 36 50
Article 100(100a) 48–50, 86
Article 130r 50
Article 130s 50
Article 130t 50
Article 213 50
Article 235 50
Asbestos 79, 87, 95
Assessment endpoints 15
Audits 40-1

Balancing-based approaches 7
Basel Convention on the Control of Transboundary Movements of Hazardous Wastes 106
BAT (best available techniques and technology) 6, 7, 98
BATNEEC 78–80, 79, 80, 87

Best available techniques and technology (BAT) 6, 7, 98
Best environmental option (BEO) 123
Best practicable environmental option (BPEO) 123
Biodiversity 14
Biological organisation, hierarchy of 16
Biomagnification 25
Biosphere 12
Black List (List I) 6, 73–8
BS 7750 40–1

Cambridge Water Co. v. *Eastern Counties Leather* 37
Canada, legislation 104
Carbon monoxide 80
Carbon tax 38
CERCLA 100–1
Charges 38
Chemical Industries Association (CIA) Responsible Care Programme 42
Chemicals
  basis of controls 5
  distribution, marketing and use 4
  legislation 4
  production statistics 1, 3
  synthetic 1, 6
Chlorofluorocarbons (CFCs) 11, 95
Chronic tests 20
Clean Air Act 39, 98–9
Clean fuel programme 99
Clean Water Act 19, 99–100
Code of Federal Regulations (CFR) 92
Comité Europeén de Normalisation (CEN) 113
Command and control regulation 34–5
Community Ecolabel Award Scheme 88

Concise International Chemical Assessment
   Documents (CICADs)   109
Consumer Product Safety Act (CPSA)   102
Consumer Protection Act 1987   58
Contamination   34
Control of Industrial Major Accident Hazards
   (CIMAH) Regulations   85
Control of Pollution Act (COPA)   57, 83
Controls
   basic questions   11–12
   classifying   5
   emissions   4, 34, 35, 39, 78–80
   instruments   33
   legislation   4
   on specific substances   84
Convention on the Prevention of Marine
   Pollution by Dumping Waste and Other
   Matter   108
COPA   57, 83
Cost-benefit approach   124–7
Council of Environmental Quality
   (CEQ)   92
Council of Europe   111, 112
Council of Ministers   46
Court of Auditors   46

Dangerous substances   68, 72–8
Decision trees   28
Decisions (EC/EU)   51
Deposit/refund schemes   39
Differential taxation   4
Directives (EC/EU)   51, 58
   67/548   59–67, 71, 111
   73/404   84
   73/405   84
   75/439   84
   75/442   84
   76/464   72–8
   76/769   81–4
   79/117   69
   80/779   80
   82/176   77
   82/242   84
   82/501   84
   82/884   80
   83/189   85
   83/513   77
   84/156   77
   84/360   78–80
   84/491   77
   85/203   80

   86/280   77
   87/18   60
   87/217   79, 87
   88/347   77
   88/379   68
   88/609   79
   90/313   88
   90/415   77
   91/414   68–9
   92/32   60
   93/112   62
   93/67   66
   94/10   85
   94/67   84

Ecolabel   41, 88
Ecolabelling Board   88
Economic instruments   38–40
Ecosystems   12, 13
   equilibrium/stability approach   15
   health   13
   services provided by   15
   stable   14
   structural aspects   19
Ecotoxicity tests   60
Ecotoxicological tests   20, 21, 25
Ecotoxicology   11
EINECS   62, 70, 72
Emission controls   4, 34, 35, 39, 78–80
Emission values   35, 98
Endocrine disruptors   120
Energy tax   38
Enforcement
   costs   7
   incentives   39–40
Environment
   adverse effects   11
   concept   12
   protection   12–15
Environment Act 1995   58
Environment Protection Act 1990   80
Environmental audits and assessments   118
Environmental concentrations, moni-
   toring   24
Environmental impact statements
   (EISs)   102
Environmental Leadership   97
Environmental management systems   40
Environmental policy, EC   52–4
Environmental Protection Act 1990   7, 36, 57,
   78, 80, 83, 87

Environmental Protection Agency
(EPA)   84, 91–2, 95, 97, 98, 100
Environmental protection legislation. See
Legislation
Environmental quality standards (EQSs)   34,
35
Equilibrium/stability approach to eco-
systems   15
European Chemicals Bureau   46
European Commission   46
co-decision procedure   49
cooperation procedure   48
European Communities Act 1972   58, 67, 72,
80, 84–5
European Community   45, 46, 112
European Court of Justice   46
European Economic Community (EEC)   45
European Environment Agency (EEA)   46,
119
European Environmental Bureau (EEB)   113
European Inventory of Existing Commercial
Chemical Substances (EINECS)   59
European List of Notified New Chemical
Substances (ELINCS)   59
European Parliament   46, 47
European Union (EU)   45, 112
institutions   46–7
legislation   57–89
Event tree   30
Existing substances   69–72, 95–7

Failure potential   30–1
Fallibility   22
Federal Environmental Pesticide Control Act
(FEPCA) 1972   97
Federal Food, Drug and Cosmetics Act
(FFDCA)   98, 101
Federal Hazardous Substances Act
(FHS)   102
Federal Insecticide, Fungicide and
Rodenticide Act (FIFRA) 1947   97–8
Federal Water Pollution Control Act 1948   99
Fifth Action Programme   86–7
Food and Agriculture Organisation of the
United Nations (FAO)   107
Food and Drug Administration (FDA)   98
Food and Environment Protection Act
1985   58
Food chain   13
Food-feeder channels   12

Freedom of access to information on the
environment   88

Gaia theory   12
Good laboratory practice (GLP)   60, 109
Greenhouse gases   17
Grey List (List II)   6, 73–8

Harmonised electronic data set
(HEDSET)   71
Hazard-based approach   6
Hazard identification   7, 22
Hazard potential   22
Hazardous Materials Transportation Act
(HMTA)   102
Health and Safety at Work, etc. Act 1974   58,
67, 80, 85
Health and Safety Executive (HSE)   67
High production volume (HPV)   27
Hydrocarbons   80, 98

Ignorance   22
Incentive schemes   39
Indicators of disturbance   18
Information-gathering legislation   4, 58–72
Institute of Freshwater Ecology   18
Integrated pollution prevention and control
(IPPC)   86–9, 122
Intergovernmental Conference (IGC) of
Member States   46
Intergovernmental Forum on Chemical
Safety   108
Intergovernmental Oceanographic
Commission (IOC)   108
International Centre on Occupational Safety
and Health Information (CIS)   107
International Code of Conduct on the
Distribution and Use of Pesticides   107
International Conference on Chemical
Safety   108
International Labour Organisation
(ILO)   107
International Maritime Consultation
Organisation (IMCO)   108
International Maritime Organisation
(IMO)   108
International organisations and
programmes   105–15
International Programme for the
Improvement of Working Conditions
and Environment (PIACT)   107

International Programme on Chemical Safety
(IPCS)   107
International Register of Potentially Toxic
Chemicals (IRPTC)   106
International Standards Organisation
(ISO)   113

Labelling   41, 59–67, 97
Laboratory tests   6
Lake sediments   18
Lead   80, 98
in petrol   4, 80
Least impact per unit function   41
Legislation
Canada   104
chemicals   4
classification   4
controlling   72–85
EC/EU   47, 51, 57–89
future challenges   121–2
information-gathering   4, 58–72
instruments   33, 51
involvement of institutions   47–51
trends   117–20
UK   57–89
US   91–104
Level of organisation of observations   16
Life-cycle assessments (LCA)   41, 88
List I (Black List)   6, 73–8
List II (Grey List)   6, 73–8
London Guidelines for the Exchange of
Information on International
Trade   106
Lowest observed effect concentration
(LOEC)   20–1

Man and Biosphere Programme (MAB)   108
Management decisions   11
Market-based approaches   99
Market (economic) instruments   4
Market forces   39
Marketing and use controls   81–4
Mathematical models   29
Measurement endpoints   15
Models   17, 18
Montreal Protocol on Substances that deplete
the Ozone Layer   106
Multivariate analysis   18

National Air Quality Strategy   80

National ambient air quality standards
(NAAQSs)   98
National Environmental Policy Act
(NEPA)   92, 102
National Environmental Policy Plan   42
National pollutant discharge elimination
system (NPDES)   99
National Priority List (NPL)   100
National Rivers Authority (NRA)   58
NEEC (not entailing excessive costs)   7
Negative warnings   41
Negligence   37
New chemicals   94–5
labelling   59–67
Nitrogen dioxide   80
Nitrogen oxides   79, 80, 98
NOEC (no observed effect concentra-
tion)   20–1, 67, 76
Non-governmental organisations
(NGOs)   112–13
North Sea Conference   111–12
Notification of Installations Handling of
Hazardous Substances (NIHHS)
Regulations 85
Notifications   60
Nuisance   36–7

Occupational Safety and Health Act
(OSHA)   101
Organic compounds   80
Organisation for Economic Cooperation and
Development (OECD)   38, 43, 71, 112,
113
Chemicals Programme   109–11
guidelines for testing chemicals   109–11
Organisms
in environment   13
interactions   12
knock-on effects for   14
within ecosystems   13
Ozone layer   11

Paris Commission (PARCOM)   111, 112
PARs   27
Particulates   80, 98
PCBs   95
PEC/PNEC ratio   67, 72
Pentachlorophenol (PCP)   50
Pesticides   59, 68–9

Petrol
   lead content in   80
   unleaded   4
pH values   17
Physico-chemical properties   24
Plants   50
Pollution control   4
   aquatic systems   72–8
   emissions   78–80
Pollution effects   34
Pollution prevention   93
Pollution Prevention Act   93
Positive labels   41
Precautionary principle   6–7
Predicted exposure concentrations
   (PECs)   22–6, 30, 67, 72
Predicted no effect concentrations
   (PNECs)   23–6, 67, 72
Predictive approach   19–22
Pre-manufacture notice (PMN)   94–5
Prior informed consent (PIC)   106
Prioritisation   7, 27–9, 71
Priority lists   71
Private action   35–8
Private nuisance   36
Public nuisance   36

QPARs   27
QSARs   27, 29, 94, 97

Ranking procedures   29–30
Red List   29
Reference sites   18
Registers   40
Regulations (EC/EU)   1, 4, 51, 59
   No.793/93   69–72, 119
   No.880/92   88
   No.1210/90   46
   No.1488/94   71
   No.1836/93   87
Regulatory impact assessment (RIA)   126–7
Resource Conservation and Recovery Act
   (RCRA)   102
Retrospective programmes   17–19
Risk analysis   7
Risk assessment   7, 19, 22–7, 66, 67, 117
   accidents   30–1, 123–4
   basic principles   23
   integrating   122

Risk-based approach   6
Risk characterisation   27, 66, 67
   basic elements   26
Risk management   33–43, 95–7
   accidents   123–4
   by command and control   34–5
   by private action   35–8
   forms   33
   options   127–9
   voluntary agreements   42–3
Risk reduction programme   43, 67
Risk sources, methods of control   34
Rylands v. Fletcher   36–7

Safe Drinking Water Act (SDWA)   101
SARA   101
SARs   27, 94
Scientific Advisory Committee on Toxicity
   and Ecotoxicity of Chemicals   76
Seventh Amendment   59
Seveso Directive   85
Sewage wastes   18
Single European Act (SEA)   45, 47
Sixth Amendment   59
Smoke   80
Species effects   15
Species identification   18
Specific substances, controls on   84
Standard operating procedures (SOPs)   109
Standards organisations   113–15
State of the environment   117–20
Statutory nuisance   36
Stochasticity   22
Suborganismic effects   16
Subsidies   39
Sulphur dioxide   39, 79, 80
Sulphur oxide   98
Suspicion-based approach   6
Sustainable development debate (SDD)   8

Tax differential   38
Taxation   4
TBT   126
Technology-based approach   6
Test organisms   20
Tort law   36, 37
Toxic endpoints   28
Toxic Release Inventory (TRI)   101
Toxic Substances Control Act (TSCA)   93–7
Toxicity assessment of complex
   mixtures   123

Toxicity identification evaluation (TIE) approaches 19
Tradable quotas 39
Treaty Establishing the European Community (EC Treaty) 45
Treaty of Rome 45, 46
Treaty on European Union (TU) 45
Trespass 37–8
TSCA 95, 96, 97
Type 1 error 24, 121
Type 2 error 24, 121

United Kingdom, legislation 57–89
United Nations Educational, Scientific and Cultural Organisation (UNESCO) 108
United Nations Environment Programme (UNEP) 106–7, 112
United Nations Industrial Development Organisation (UNIDO) 108
United Nations programmes 105–9
Unleaded petrol 4
US Environmental Protection Agency (USEPA) 19

US legislation 91–104
US Toxic Release Inventory (TRI) 40
US Toxic Substances Control Act 27

Vienna Convention for the Protection of the Ozone Layer 106
VOCs 80
Voluntary agreements 42–3

Waste controls 84
Water Industry Act 1991 58, 78
Water pollution 99–100
Water Resources Act 1991 58, 78
Wildlife and Wilderness Acts 103
World at a Glance (WAAG) report on international chemicals programmes 104
World Bank 108
World Health Organisation (WHO) 107
World Wide Fund for Nature (WWF) 113

Index compiled by Geoffrey C. Jones